WHAT
MAKES
OLGA
RUN?

WHAT MAKES OLGA RUN?

The Mystery of the 90-Something Track Star, and What She Can Teach Us About Living Longer, Happier Lives

BRUCE GRIERSON

RANDOM HOUSE CANADA

PUBLISHED BY RANDOM HOUSE CANADA

Copyright © 2014 by Bruce Grierson

www.randomhouse.ca

Random House Canada and colophon are registered trademarks.

Library and Archives Canada Cataloguing in Publication

Grierson, Bruce, author

What makes Olga run? : the mystery of the 90-something track star, and what she can teach us about living longer, happier lives / Bruce Grierson.

Includes bibliographical references and index.

Issued in print and electronic formats.

ISBN 978-0-307-36345-9

1. Kotelko, Olga, 1919–. 2. Kotelko, Olga, 1919– —Health. 3. Track and field athletes—Canada—Biography. 4. Sports for older people—Physiological aspects. I. Title.

GV1060.72.K68G75 2013   796.42092   C2013-900757-1

Text design by Meryl Sussman Levavi
Cover design by Kelly Hill
Cover image: © Patrik Giardino / The New York Times Magazine

Printed and bound in the United States of America

10 9 8 7 6 5 4 3 2

To Jennifer: For the Long Run

"The water doesn't know how old you are."
—DARA TORRES

CONTENTS

WHAT MAKES OLGA RUN?

PROLOGUE

EVER SINCE SHE turned 90, Olga Kotelko has presented a problem for organizers of the track meets she enters: Whom does she compete against?

The issue surfaced prominently in the 60-meter-dash final at the World Masters Indoor Athletics Championships, in Kamloops, British Columbia, in 2010. Olga found herself, well, in a class by herself. There just aren't many nonagenarian sprinters—even when you draw from the whole planet. The next-oldest woman in this meet, Californian Johnnye Vallien, was 84.

So there Olga was, 91, bespandexed and elfin, lumped in with the men.

In lane one stood Orville Rogers, 91, a long-striding retired Braniff Airways pilot and the world-record holder in the mile for men over 90. Next to him: Belgian Emiel Pauwels, 90, another world-record middle-distance man (1,500-meter), in bright orange track spikes, who would later make everyone nervous as he ran most of the 3,000-meter final with his left shoelace untied. Front and center: Ugo Sansonetti, 92, a former frozen-food magnate from Rome, in a blue sleeveless skinsuit, his tanned biceps bulging like small baked potatoes.

Olga drew the inside lane, rounding out the field. She wore black tights and a long-sleeved white shirt—the modest uniform she wears no matter the weather.

She'd been worried about her start. She's not a good starter. She can get rattled. Sometimes, when the gun sounds or even a fraction of a second before, she takes a step *backward*. But today she started clean and mechanically strong, piston-pumping her arms, generating enough wind to pin her hair back a bit.

It's no longer strange to see geriatric runners: every big-city marathon has its share of valiant, white-haired competitors who spark bursts of applause as they shuffle past. But it *is* strange to watch 90-year-olds sprint. Kids and dogs and young adults run full-out. But old folks? The incongruity of that image inspired a television commercial that Ugo Sansonetti shot for Bertolli margarine not long ago. A runaway baby carriage is seen careening through the streets of Rome, until a fissure-faced old bystander—Ugo—springs into action and chases it down.

Sansonetti crossed the line first at Kamloops, in a world-record time of 11.57 seconds. He bounced around in the run-off area, arms overhead in triumph, as Rogers glided in behind him at 12.82. Olga came third, at just over 15 seconds. She looked concerned for Pauwels, the Belgian, who had caught a spike and crashed down hard, then picked himself up and limped in last.

She was cool with running against the guys. "That one fellow was pretty fast," she said, on the way to the changing area. She had gotten used to this. When you're the fastest 91-year-old woman on the planet, either you compete against younger women or you run against the guys.

Just how good is Olga? There are a couple of ways to put her in perspective.

She currently holds twenty-six world records. She set twenty world records in a single year, 2009. She hits these totals in part by entering more events than everybody else, including a couple that nobody else in the world her age attempts. She will often do six throwing events, three sprints, and three jumps. (At age 88, she considered adding the pole vault, but was deterred by practical considerations. "What do you do with the pole—strap it to the roof of the car? Check it on to the plane?")

Track records, at the elite level, tend to fall by fingernail parings of time and distance: fractions of seconds, portions of inches. At the 2009 World Masters Athletics Championships in Lahti, Finland, Olga threw a javelin almost *twenty feet* farther than her nearest rival. At the World Masters Games in Sydney, Australia, in 2009, Olga's time in the 100-meter dash—23.95 seconds—would have won the women's 80–84 division—two age brackets down.

Olga stands five feet and a half an inch. She weighs 130 pounds. For her size—and this may be the most curious thing about her—she has extraordinary power. It can be surprising, after her slo-mo windup, to see how far the things she throws go.

On the hammer throw pitch in Kamloops, she took her place with the other competitors. Big guys with leather gloves paced around, shaking their hands out. Olga removed her glasses. There was a sudden and brief sense of menace; when a little old lady starts swinging a three-pound cannonball around her head, a good outcome is not guaranteed. But the thing sailed, straight and true. "If I spun I could throw it farther," Olga says, but after watching somebody very old fall that way she has decided not to risk it.

Olga got more leg into the second throw. But the trajectory wasn't what she liked. She made a little swan's head gesture with her hand, to remind herself. Routinely, Olga performs better on every subsequent attempt as she recalibrates and tries again. It's like watching a marksman bracket the bull's-eye and then draw in: 12.72 meters. 13.37. 13.92. In ten minutes she added four feet of distance. "New world record," a disembodied voice said over the loudspeaker.

There is a formula called "age-grading" that's used to put the performances of older athletes in perspective. Age-graded scores tell us how impressed we should be by what a masters athlete—placed in categories from ages 35 to 105—just did. A set of tables plots a given performance against the expected decline of the human body, and expresses it as a percentage. So, theoretically, 100 percent is the high-water mark for a human being of that age.

But a number of Olga's marks—in shot put, high jump, 100-meter dash—top 100 percent. In Sydney she threw the shot put 5.6 meters—which age-grades out at 119 percent. If you plug Olga's 23.95 100-meter-dash time from Sydney into the tables, you find it's exactly equivalent to American Olympian Florence Griffith Joyner's prevailing, suspicious, and thought-to-be-untouchable world record of 10.49 seconds.

Remarkably, when you age-grade, you find Olga is not only holding her own but in some cases getting better—which suggests that either the tables are wonky or Olga is. Most likely both are true. "She throws off the curve, because she's doing things nobody's ever done," says Ken Stone, editor of masters-track.com, the watering hole of the masters track community. (Motto: "Older, Slower, Lower.") But at the same time, some recent performances speak for themselves. "I threw the hammer farther this year than I did two years ago," she mentioned recently, offhandedly. "How do you explain that?" No two ways about it: Olga is defying, or rewriting, our understanding of the retention of human physical capability.

When people hear how old she is, they seem to look at her more deeply, at her face. To be blunt: she is not aging normally.

"How old do you *feel*?" I asked her on her 91st birthday.

She thought about that. "Fifty?" She gave a half shrug. "I still have the energy I had at fifty," she said. "More. Where is it coming from? Honestly, I don't know. I wish I knew. It's a mystery even to me."

THEY say she is like Grandma Moses, in the sense that she found her calling very late in life. But while Grandma Moses took up painting out of desperation, to make ends meet, Olga took up track, at age 77, for fun. A dozen years retired from her career as an elementary school teacher, she still had lightning in her that needed grounding.

For the first half of her track career—till around age 85—she coasted under the radar, quietly breaking world records within a subculture obscure enough (track and field for old people!) that many people still don't know it exists. But since she turned 90, media interest in Olga has made it hard for her to hide. Reporters from around the world have made pilgrimages to her home on the flank of Hollyburn Mountain, overlooking the Pacific Ocean in West Vancouver. There is usually pleasant chit-chat, and the reporter, having heard the jokes that the fountain of youth burbles up in her backyard, just behind the organic vegetable garden, makes an excuse to snoop around.

Because, seriously: when you're breaking records, rather than hips, at an age most people will never live to see . . . what gives?

Aging is supposed to be one of those nonnegotiables: just the cost of doing life. It is, as the Stanford professor of medicine Walter Bortz put it, "a stern expression of the Second Law of Thermodynamics from which there is no respite." So how do you begin to think about someone like Olga, short of demanding to see her birth certificate or otherwise trying to debunk her story? What seems to be required is a leap in the way we understand the aging process, as scientific advancements slowly reveal clues.

Indeed, for a stage of life that 90 million baby boomers in North America alone are barreling toward, surprisingly little is known about old age. (Or maybe it's not so surprising: old age was a nonissue in human history until very recently. More than half the people *who have ever* reached age 65 are alive today, some demographers believe.)

But lack of data has never stood in the way of forceful opinion. Just about everyone has a guess about what makes Olga run. It's genes. It's diet. It's temperament. It's her unusual paleo sleep patterns. It's the fighting spirit of the Cossack general from whom she is descended. It's energy: she is vibrating at a higher frequency than most anyone else. It's the miracle of exercise itself, compounded over a long, long lifetime. It's performance-enhancing drugs. (Actually, we can rule that one out: she's clean.)

In the morning—very early or merely early, depending on the day—Olga gets up and puts the kettle on for Krakus, a Polish coffee substitute made from roasted flax, barley, and beetroot. Then she heads into the bathroom to wash her face. The woman looking back at her in the mirror contains multitudes. She is the sweet-natured grandma with a competitive streak so ferocious she aims to set marks that eclipse the past greats and make up-and-comers think about maybe trying out scrapbooking instead. The farm girl who outlived ten siblings and one of her two daughters, and ponders what is being asked of her, the survivor. The levelheaded pragmatist with abiding faith in water, reflexology, and a curious routine of massaging and stretching her muscles in the dead of night.

Olga and I met four years ago. I was one of those nosy writers who showed up at her home: the in-law apartment in the basement of her daughter and son-in-law's house.

It was a tidy space. Her own paintings of landscapes, marinas, and flowers from the garden hung on the walls. She offered tea. She had kind eyes and carefully applied lipstick and a shy demeanor. It was hard to reconcile this gentle person with the figure conjured by the growing mythology—a woman whose warrior fire is so intense that when she first announced she was interested in track, and a well-meaning friend said, "The event you want is the racewalk," the advice literally did not make sense to her. Seriously? Walking or racing: which were they talking about here?

I'd hoped we could go to the running oval at the local high school so I could observe Olga in action. But it was pounding rain outside, so instead we stayed in and talked. We went through her notebooks documenting thirteen years of increasingly unlikely levels of achievement in track and field. At the end she cracked open the door to a closet. It looked like what Geraldo had hoped to find in Al Capone's vault. Gold shone forth. Hundreds of medals hung from groaning hooks.

If a writer can become infected by a story, that's what happened to me that day. Olga started to feel more interesting, more *important* than everything else I was thinking about. She tolerated my endless questions, probably because they were the same kinds of questions she had about herself. Eventually we struck a deal: we would become a team. We would explore the mystery of her together. She would offer herself up to science while I took notes. We would tap the expertise of some of the top practitioners of exercise physiology and gerontology, neuropsychology and evolutionary medicine.

To be honest, I had a secondary, selfish motive. I felt a personal stake in what the data about Olga would reveal, because I had my own mystery to solve.

You see, I used to be fit, too—not Olga-fit, but fit. I started exercising, in a pretty committed and deliberate way, when I was a skinny kid of nine. I got good tennis coaching and started beating adults, via a style of steady, craven retrieval that wore them out (and ticked them off). A running addiction that took hold around age twenty ensured that I kept grinding out workouts into my thirties and forties. Pairs of Nikes—each dutifully retired after six hundred miles—piled up in the closet like cordwood until:

Boom. Something . . . gave. Age flooded in all at once. It really did feel that sudden. Gone, just like that, the jump, the stamina, the drive, the memory, the hair. Gone, any right to use the noun "performance," in any context. And with those changes came a fairly radical shift in attitude. Each sunrise no longer dialed up a sense of hope but resignation.

How does this happen? I was 47. Could it be the inevitable midlife swoon we're warned about, the one that starts with stretch jeans and topical ibuprofen and ends up with saltless dinners in the extended-care wing? Well, clearly it's not inevitable, because there before me was Olga. Whatever was happening with her was the opposite of what was happening to me. If she was the paragon of healthy aging, I was a proxy for every once-hale boomer now alarmingly on the skids.

Some serious sleuthing was in order. It would require going back in time to peek in on Olga as her current self was still under construction. What was she doing, at my age and younger, that was so forward-thinking, so right (or maybe just so lucky)? And could those habits be wrenches to fix the mistakes the rest of us are making, before it's too late?

1

Rust Never Sleeps

To grasp why Olga is apparently aging more slowly than usual, it helps to think hard about what aging actually is—and why it happens. After all, we need to die, but it's by no means obvious why we need to *get dead*, incrementally, the way we do.

Turns out, aging is a tricky thing to study. Scientists have taken several approaches.

The best one is just to wait. Start a study with kids or young adults and then follow them through the journey, quizzing and poking them every few years and record the changes as they approach their end. And indeed, we now have longitudinal data from people over seven decades of their life.

If researchers can't wait, they look for ways to cheat by speeding up time. They observe the body in extreme environments that produce symptoms that mimic age-related decline. Scientists have trekked to Everest base camp and lower-earth orbit, and watched hypoxia and weightlessness do its corrosive work on the cells and bones of climbers and astronauts. They've studied people suffering diseases whose symptoms look like time-lapse aging.

And what, after all this spadework, can we conclude?

There are competing theories about what's actually happening when we age.

We're rusting. Humdrum daily metabolism generates cellular garbage until the janitorial staff can't mop it up fast enough. Oxidative damage ensues, along with inflammation in every cell, until something important fails or a chronic disease emerges.

We're timing out. Nature planned for our obsolescence. It put a limit on how many times our cells can safely divide. After reaching that number our chromosomes start sticking together, DNA is damaged, and the body sends a suicide order to put an end to the whole failing enterprise, right on cue.

We're borrowing from our future to pay for our youth. Aging, in this scheme, is a devil's bargain. Some genes that help us to be strong and fertile in early life turn destructive beyond our reproductive years, when nature decides we are now of best use as mulch.

There are other theories, too. None of them has been scientifically proven. All we know is that we slide gradually downhill, and it's a lot easier to speed up the descent than to slow it down. But some people—such as Olga—seem to have acquired better brakes. Something is helping Olga. Something is protecting her mitochondria or her chromosome-shielding telomeres, or boosting her immune system, or suppressing inflammation, or repairing DNA mutations, or keeping her cell membranes supple. Perhaps part of what's protecting her is her own mind. That is, she is refusing to become a wizened little old lady at the normal rate because she simply does not believe she is one.

The Wall

I have heard grumbling from other athletes at masters track meets when the cameras follow the very old participants around. A reporter pursuing Olga will walk right past jocks in their 60s and 70s who are world champions in their own right. Why should so much fuss be made over the 90-something woman who can't throw the javelin as far they do, or run the 200-meter dash nearly as fast?

Because the 90-something woman is doing something different. It is the same, but it is different, because *she* is different. She is on the other side of a wall that the 70-year-old knows nothing about.

On a fall morning not long ago, muscle physiologist Russell Hepple stood before a group of McGill University kinesiology students. The class was K-201, "Physical Activity in Aging." Hepple, who is nearing 50 and has groovy sideburns and is still built like the competitive runner he used to be, was guest-lecturing for the regular prof, Tanja Taivassalo—who happens to be his wife. He had a few surprises in store, including a guest from out west.

Hepple put up a picture on the overhead screen: a buff, shirtless man holding barbells. A sidelight defined the man's muscles like the sun setting over dunes.

"How old do you think this guy is?" Hepple asked. "Some of you are thinking—can you give me his phone number? Obviously he's not twenty-five. His shorts are far too high."

The students quietly conferred.

"Give up? He's sixty-seven. But if you didn't see his head you might be thinking twenty-five or thirty. Sixty-seven. Pretty damn impressive."

As it happens, the photographer caught up with the same man a dozen years later.

"And," said Hepple, "here he is."

Now 79, the man duplicated his pose from the previous picture. "He's been carrying around the same dumbbells for twelve years," Hepple said. "Wearing pretty much the same gym shorts, at the same altitude."

But he had changed. He'd deflated. "He's probably lost thirty to forty percent of his muscle mass," Hepple said.

"This tells me two things. One, something about this transition into the seventh and eighth decade of life is very important. Two, even if you continue to exercise at a very high level, you can't sustain."

Why? "It's a question we don't have a firm answer to yet. But obviously we're working on it."

The effects of aging are every bit as obvious in runners. When Pixar was making the film *Up,* the animators realized they needed to depict an old guy's rejuvenation of spirit. But they had to be careful not to make Carl Fredricksen look *too* spry, or he would no longer look old. For a model of an old guy running as fast as an old guy can run, they viewed clips of competitors at the National Senior Games.

Hepple put up a chart of world-record sprint times by age group. The line slanted upward as older runners posted slower times. Around age 60 the slope started steepening. And then in the late 70s, it shot up. Both men and women suffered the same fate. Good-bye muscle mass, good-bye performance.

In the front row sat Olga, taking notes. In a moment Hepple would call her up to speak to the class. Here was the exception to prove the rule, a genuine scientific curiosity who somewhat defied the expected rate of decline. If we could understand what's going on with Olga it might help us understand aging. Hepple and his wife Taivassalo, a cell-mitochondria specialist, had flown Olga to Montreal that week for extensive muscle testing.

The health of muscle, Hepple said, tells us a lot about the health of the body more generally. "Things like cognitive function and renal function, cardiac function, cardiovascular function—all these probably have similar trajectories of decline." In other words, a lot that we couldn't see—not just his obviously shrinking muscle—was starting to fail in the gentleman in the high-waisted shorts. From the quality of Olga's muscle, the scientists could likewise make inferences about how her other systems were holding up.

After stepping up to share her story with the McGill class, Olga fielded questions. The students were curious about the usual things: diet and training and sleep. When, someone wanted to know, did she realize she was different?

Born This Way?

January 1997. Olga crossed the rubber oval at the University of Arizona at Tucson and headed for the infield grass. She wore bib 777: a good omen, she thought. But she had butterflies. It was her first international meet—really her first proper test of whether she was any good at this new sport she'd taken up just a few months before.

Since retiring from a thirty-four-year teaching career, Olga had looked for competitive outlets to fill her days. She'd joined a softball league at age 70—"Slo-Pitch," an at-first-annoying variant of the baseball she knew so well from childhood. (Annoying because she grew restless standing at the plate waiting for the ball to approach in its lazy arc.) She picked it up quickly. She played five positions. That lasted until the day she was plowed down in the outfield by an overzealous teammate chasing the same pop fly.

Track and field felt like a safer option and a logical sidestep. She already had the wheels—she'd been a good base runner. She had the arm—she could throw runners out on one hop from the shallow outfield. She just needed to learn the mechanics of these new events. She went to the library and took out a few books. A local coach gave her some instruction and put in her hands the tools of the trade.

And now here she was in Tucson. The other women in the meet all knew one another, but Olga knew no one and no one knew her.

She had entered three events, which, looking back, she considers folly. ("Can you imagine spending all that money for just three events?" she says.) One was the javelin.

The javelin she'd been throwing back home weighed six hundred grams. She thought that was standard. But now a field judge in a white polo shirt handed her the javelin she'd be throwing today. It seemed way lighter. It *was* lighter: just four hundred grams. Also shorter. It felt like a practice javelin—and her first few throws were erratic as she adjusted to the change.

On her fifth throw Olga launched a missile.

"Ladies," the judge told the ten other competitors in the 75-plus group, "you have fifty-seven feet to beat."

There was a stunned silence. Where did that come from? Who *was* this person?

The other women lifted their games. They began to close the gap: 52 feet, 53. Then Olga stepped up for her final throw. With everything that was in her she let the spear fly. "It felt," she recalls, "like an angel carried it away."

"Sixty-five feet six inches," the judge said.

That settled things.

"The other women went wild," Olga recalls. "They started asking me questions. What do you eat? Do you have a trainer?" (A thought occurred to Olga too impolite to share: *If I'm throwing four meters farther than my opponents, do I need a trainer?*) If her spear had had a lightning bolt on it, like Robert Redford's bat in that mystical baseball movie, no one would have been too surprised. For if ever there was a "natural," it was Olga. She must have been born this way, blessed with this.

Right? Isn't that so often the assumption? In the face of the mystery of extreme performance—just as with the mystery of longevity—it's comforting to have a one-word answer: genes.

Not long ago, Dr. Nir Barzilai, the director of the Institute for Aging Research at Albert Einstein College of Medicine in New York, published a study that guided our thinking along those lines. Many of the centenarians he had tested achieved super old age despite some dubious lifestyle habits. A number were big smokers or drinkers or unhealthily overweight. "But with the right genes," he told a reporter, "their bodies are protected." Olga, you'd think, might likewise be benefiting from some freakishly robust DNA that explains everything.

With that assumption on the table, we set out on our first quest.

The "Super Senior" Advantage

On a bright May day in 2012, Olga and I pulled up in front of the BC Cancer Agency in Vancouver. "I'm quite familiar with this place," Olga said. In the late 1990s, she used to drive Nadine, the eldest of her two daughters, here for treatment, after Nadine was diagnosed with non-Hodgkin's lymphoma at age 53. Nadine was a schoolteacher like her mom. One of the world's best specialists in that discipline oversaw Nadine's care, but he couldn't save her. Only recently has Olga been able to come back to this neighborhood without difficulty. "Life goes on," she said as the doors whispered open.

Today's task was happier. Olga was participating in geneticist Angela Brooks-Wilson's Healthy Aging Study—an investigation of more than five hundred people dubbed "Super Seniors."

If you're a Super Senior, you are over 85 with a clean bill of health. You have run between the raindrops, diseasewise. You have escaped the "Big Five" killers: cancer, cardiovascular disease, Alzheimer's, diabetes, and pulmonary disease. You're rare: only about 2 percent of all 85-year-olds can make this claim. (The average 75-year-old suffers from three chronic medical conditions, according to the Centers for Disease Control and Prevention.) But having reached this point, paradoxically, your odds of staying free and clear are good. Beyond this age the death rate plateaus and actually starts declining—possibly because the people left standing are the cream of the species.

Brooks-Wilson and study coordinator Johanna Schuetz toured us around the facility. Olga took Schuetz's arm, not for support but for connection. We passed grad students loading centrifuges. Big white freezers were jammed into every nook. The freezers hold ten thousand or so DNA samples, which will remain on ice until questions scientists never thought to ask suddenly seem like promising leads.

"Your DNA is, by definition, a good sequence," Brooks-Wilson told Olga. "Whatever you have is compatible with a healthy life." Olga's genetic material would be compared to that of a randomly picked control group of people around my age. People who are, as Brooks-Wilson puts it, "at that stage of life before the tough stuff starts."

The analysis will zero in on about 1 percent of Olga's genome—carefully selected parts of the DNA thought to be functionally important. Those million or so "exons" tell a kind of CliffsNotes version of a person's genetic story. Computers in the Cancer Agency will look for patterns of single nucleotide polymorphisms, or SNPs (pronounced "snips")—small genetic changes—that occur more frequently in Super Seniors, and so may confer some protective advantage. (In the very, *very* old, the role of genetics rises. Scientists connected with the New England Centenarian Study now claim they can predict if a kid will live to age 100 from his DNA sample alone, with 61 percent accuracy.) And while very few traits or diseases can be explained by the presence or absence of a particular SNP, some can, and are.

"We think longevity is probably seventy to seventy-five percent lifestyle," Brooks-Wilson said. That benchmark number comes from a study of nearly three thousand Danish twins drawn from the general population. That means, for the vast majority of us, roughly a quarter of healthy aging is about the protection you luckily inherited, and three-quarters is how you played the hand you were dealt.

People who roar into their 90s and 100s very often had long-lived parents. (When Brooks-Wilson surveyed the ages of the parents of her Super Seniors, she found they lived *fifteen years* longer than average.) But Olga doesn't necessarily have that pedigree. Her mother, Anna, did live to a ripe old age: she died at 85 after a bout of pneumonia. But her father, Wasyl, died at 74 (also of pneumonia). Her maternal granddad, Michaylo, died at 82. Her paternal grandfather, Stephan, was 79. Her maternal grandmother, Hafia, was 63. Her paternal grandmother, Anastazia, was 62. Among her siblings Olga is by far the longest lived. She is the last alive among eleven kids, even though she was tucked in the middle, with four older siblings. None of the siblings was remotely as active in adulthood as she.

Olga's genome could reveal secrets that will benefit others down the road, in ways that are hard to predict.

In the meantime, we decided to throw our curiosity a bone. We sent for a couple of kits from a consumer genomics lab— one for Olga and one for me. When they arrived, we met for lunch in the cafeteria of the seniors' center near her home. I ripped open the packaging and placed on the table two vials for saliva samples.

The human genome was not part of the curriculum when Olga started teaching science in the 1940s. Body science was lumped into the hopper of "health." Back then, the nature

versus nurture debate was simpler. Heritable traits largely determined who we are: that's Olga's remembrance of the received
wisdom. Now we appreciate how genes and environment are
complicatedly intertwined. Our DNA is not so much a blueprint
as a starting point; it contains a staggering number of switches
that could go either way. *Everything significant that happens to us
potentially alters gene expression*: that claim becomes more defen-
sible by the day. If it proves true, then meeting Olga has rede-
fined me.

Genetic profiling thus must be taken with a pretty big grain
of salt. Even when a scenario screams "heritability," the picture
is generally more complicated. In his book *The Sports Gene*,
David Epstein tells of a group of elite marathoners from Kenya's
Rift Valley—a cohort so gifted that it can be discouraging for
other runners to face them in meets. The feeling among some
is that there's just no beating the Kalenjin aces, no matter how
hard you train.

But while it's true that the Kalenjin have physiological advantages that make them ideally suited to distance running—
including ultralight, long-limbed frames that shed heat, narrow
pelvises that reduce compression of the hip joints, and super-
skinny lower legs with calves snugged up close to the knee,
which amounts to real energy savings over the long haul—it's
not fair to chalk up their dominance solely to superior DNA. The
Kalenjin are deeply committed to distance running, almost from
the time they can walk. And the elites have a work ethic few
rivals can match. Moreover, inside those thoroughbred bodies,
attributes that look like genetic supercharging are in fact environmental adaptations. The thin air of the Kenyan highlands
triggers the making of more red blood cells to ship oxygen to the
muscles. The world-beating Kalenjins, then, are a paean to
nature *and* nurture, "a combination of physiological advantages
and a unique crucible in which to develop them," says Epstein.

Genetic testing can't predict who among us has the royal jelly to "be all we can be." The best it can do is spell out your destiny in terms of probabilities. Certainty is a mirage. Then again, we weren't looking for certainty: just clues.

Olga popped the top on her vial. At the next table, a gentleman in a tweed jacket, perhaps fearing the government was conducting some shadowy eugenics project, watched with curiosity bordering on alarm. Then we packed up our captured saliva and sent it off to California.

Superstar Genes, Warrior Genes, and Probabilities

At least 150 genes are known to be linked to athletic performance. Some traits, scientists have recently discovered, are under the control of a single gene. For example, researchers have found a gene variant in mice that supercharges the muscle with mitochondria, allowing the mice to run and run. Does Olga have that variant, too? Hard to say: scientists haven't yet found it in humans. But she probably doesn't, given that endurance is not her strong suit.

One gene that does seem to work alone is the so-called superstar gene, ACTN3, on chromosome eleven. It promotes either muscular power or muscular endurance, depending on which version of it you get: the fast or the slow. Each of us is born with two copies of ACTN3. Around one in three people of European stock get two copies of the fast version—the "on" switch for jumping and sprinting. Around one in five get two copies of the slow version—endurance. Around half of us get one of each. Virtually every Olympic sprinter ever tested has had two copies of fast-ACTN3—the speed-and-power allele.

It would be shocking if Olga, who is all about explosive power, didn't have two "on" switches—two copies of fast-ACTN3. And, sure enough, she does.

This was the first of the "health markers" we looked at when her results arrived.

We laid our profiles side by side. The data were grouped under categories such as "health risks," "inherited conditions," "traits," and "ancestry composition." Each score was attached to a "confidence" level, from one to four stars, that this outcome was indeed part of this individual's story. Comparing the quirky things was fun; comparing the significant ones, bracing.

Turns out I probably inherited a sweet tooth, and she didn't.

I am probably less sensitive to pain.

I am closer to a Neanderthal than she is. I have 2.9 percent Neanderthal DNA, while she has 2.5 percent. (Sounds trivial, but my score puts me in the 91st percentile of all those tested—among the knuckle draggers—and hers puts her in the 45th.)

Olga looks to be the brighter of the two of us. People with her genetic markers tend to score higher on all measures of intelligence covered by this test, including episodic memory and nonverbal intelligence. She appears to be gifted with what must be one of the absolute denominators for evolutionary success: "the ability to learn from mistakes."

How we learn is thought to involve a "reward circuit" featuring the neurotransmitter dopamine, which regulates the rate at which neurons fire. Around five years ago, scientists identified a SNP on the eleventh chromosome linked to how we respond to criticism or failure or punishment. People with a double copy of this SNP tend to be highly sensitive to negative feedback, and hence acutely motivated to avoid it in future (by "correcting" the behavior). Olga has the double copy. I don't, which makes me "much less efficient at learning to avoid errors," according to the report.

Dopamine also figures in another well-studied gene called COMT, on chromosome twenty-two. It has become known as the "worrier/warrior" gene. The COMT gene instructs the body to make an enzyme that flushes dopamine from the frontal cortex. The worrier removes it slowly, and the warrior clears it quickly. What that means, effectively, is that worriers, with more dopamine keeping the "executive function" part of their brain humming with neuronal activity, are sound planners and careful problem solvers. Warriors are more impulsive and action-oriented.

Each has its advantages. Worriers outperform warriors on a lot of cognitive tasks. But crank up the pressure and warriors tend to thrive, while worriers wilt. Also, warriors seem to be compensated for their lesser contemplative horsepower early in life by becoming slightly more likely to hang on to their marbles in later life. So if you wanted to engineer a person combining the best qualities of both alleles, you'd give them one copy of each: half worrier, half warrior.

Olga has one copy of each.

(Me, I'm double warrior. Ready, fire, aim.)

All told, Olga does have a pretty nice genetic profile. It's probably helping her cognition—both offensively (sheer computing power) and defensively (staving off dementia). She's probably getting some protection from type 1 diabetes, rheumatoid arthritis, and certain types of cancer: lung, stomach, breast. (Again, it's hard to say for certain that she's protected; we know only that people with her genetic markers are statistically less likely to get these diseases.) People with her genotype get coronary disease around half as frequently as average. With her genes and her exercise regime, she has pretty much taken this one of the Big Five killers off the table.

But her profile is by no means perfect.

Some people carry a gene variant that all but prevents Alzheimer's. Olga doesn't.

Some people have genes that make them lose weight easily on a workout regime. Olga doesn't.

Some people carry genes that make them highly responsive to aerobic exercise. Olga doesn't.

Some people carry genes that help them maximally benefit from a healthy Mediterranean diet. Olga doesn't.

Around 3 percent of us carry a gene variant that lets us get by perfectly well on much less sleep than average—leaving more waking hours to develop our talents and fulfill our potential. Olga doesn't.

She is more vulnerable than most to colorectal cancer.

As for longevity itself, it's hard to draw conclusions from a hot-spot profile. We're many years from understanding how genes interact with one another and with the environment to produce long life. But around a decade ago, scientists did find a single SNP linked to longevity. Geneticists at the Institute for Aging Research in New York studied a population of very old Ashkenazi Jews (aged 95 to 107) and found many had two copies of the variant—which possibly protects against cardiovascular disease.

Olga doesn't have it. (Neither do I.)

Of course, it's possible that Olga is getting help from genes not yet identified. Variants could be protecting her in a complicated way from injury. Others could be shaping her performance under stress. Still others could give her the intrinsic motivation to exercise in the first place. "Recent studies have shown that centenarians have just as many disease variants as ordinary people," says Brooks-Wilson. In other words: they carry the same poisons, so it's highly likely they carry some hidden antidotes, too—rare mutations that keep them going on and on.

But based on what we know now, it's far too simple to explain Olga away as a "genetic freak."

We have to look at the ways she has deployed those genes out in the world.

2

The Adversity Hypothesis

"Someday This Pain Will Be Useful to You."

IN THE STUMP of an old tree you can read the story of its life. Where the rings are widely spaced, conditions were favorable those years and it grew like gangbusters. Where the rings are tight, those were discouraging years—full of drought or unseasonable cold or stressful crowding from neighboring trees.

If Olga were a tree and you cross-sectioned her—practically the only thing scientists *haven't* done to her thus far—the spacing of her rings would likewise tell her story.

From the center the rings fan out widely, for her early years were rich and stimulating and happy.

Officially, she was born in tiny Vonda, Saskatchewan (population three hundred), in the Canadian heartland. But the 160-acre family homestead is actually closer to even tinier Smuts (population "Who's gone to Saskatoon today?"). A tongue of flatland snugged against a hillside, it had been gifted to her father as part of a government incentive program to draw settlers from eastern Europe. It was mostly stones and trees.

As a child, Olga was acutely aware of the seasons—not by changes in temperature or sunlight, but because in winter,

Mom and Dad slowed down. They were around more. On winter evenings her father, Wasyl Shawaga, mustered his eleven kids around the woodstove. He read them the classics of Ukrainian literature: the detective novels of Ivan Franco, the poems of Taras Shevchenko ("We shall . . . take our rest together . . . / And your sister-stars, meanwhile / The ageless ones, will start to shine"). Sometimes a Ukrainian translation of the Bible. Never anything by the hated Russians.

The deep emotional payoff of those moments, Olga is pretty sure, is what has kept the Ukrainian alive in her to this day. "When I get sick and can't do track anymore—and believe me I'm not going to stop any other way—I have this to look forward to," she says, of revisiting those old classics.

Her mother, Anna Bayda—a direct descendant of sixteenth-century prince Dmytro Bayda Vyshnevetsky, the Cossack folk hero who fought the Tartars and the Turks—came to Canada from the Ukraine at age 12 on the steamship *Bulgaria*. She was taciturn like her husband, and stoic. "She never complained, about anything, ever," Olga recalls of her mother, who in photographs looks exactly like Olga, only rounder.

Grandparents from both sides lived with them on the farm. Olga was closest to her maternal granddad Michaylo, who had been a policeman in the Ukraine. She liked to watch him and sketch him in pencil. She captured his portrait one day as he reclined with his feet in the oven to keep them warm. When he died his body was placed in a homemade casket in the parlor, so that everyone could pass a day of quiet observance with him. There was no formaldehyde. Olga remembers watching as liquid beaded on the bottom edge of the casket, and her mother put a little bucket down to catch the drips.

Her family weathered the Depression better than most, even as drought ravaged the crops. (At a family reunion, her brother Steve talked of starving horses that had to be put down, but

Olga doesn't remember any mercy killing. Her parents did a good job of shielding the younger kids from the hardship.)

The rings remain wide as Olga ventured out and landed her first jobs—a short stint as a secretary at a coal company and then, at age 22, teaching grades one through ten at a one-room rural schoolhouse, for a beginning salary of $700 a year. She was crazily outnumbered, but happy.

And then the rings begin tightening.

At a dance she met a handsome young insurance salesman named John Kotelko. He played the fiddle and could charm an owl out of a tree. They soon married. But within weeks of the vows, she knew something was wrong. If he ever genuinely cared about anyone but John Kotelko, that care was gone; if it had been an act, he had dropped it.

Many nights he didn't come home. He turned out to be a philanderer. And an alcoholic. He grew abusive, and yet she hung in there with him for a decade. To daughter Nadine, born in the first year of marriage when things weren't yet too bad, her father was mostly a stranger, a ghost.

One night, during a drunken rage, John produced a knife. Olga can still bring back the feeling of the blade against her throat. "If I hadn't got out of there, I would have been dead," she says. "I'm positive of it."

This was the 1950s on the Canadian prairie. A woman did not leave her husband. Olga had never even met another woman who had done so. Nevertheless, Olga, pregnant with her second daughter, gathered up 8-year-old Nadine and their minimal belongings and fled into the night. They boarded a train heading west. It was March 1953. "As far as I knew, I was the first single mom in the history of the world," Olga says.

She moved in with her sister Jean, in New Westminster, British Columbia, a bedroom community of Vancouver. She gave birth to her second daughter, Lynda. She found a job at a creosote factory down on the docks. At night she took university classes to complete her bachelor of education degree. It was a lot to manage, and she briefly became a smoker during this spell. She allowed herself two cigarettes a day. Four years later, at age 39, she quit.

But the tree rings soon widen again. The three girls were without a man in their life, but this was no problem. One day Olga decided to have a new house built on the site of their old one. Simple: you call a company, it sends out some guys and heavy equipment, no? No. "She and Nadine tore it down, all by themselves, with hand tools," Lynda recalls. In 1965, long since settled in her new home, Olga got a call that John had died, at age 43, of lung disease from a lifetime of smoking.

For decades the tree flourished. Then in the 1990s, tightness returned as Nadine got sick and eventually died in 1999.

PSYCHOLOGISTS discuss a theory known as the Adversity Hypothesis, which goes roughly like this: resilience is learned. It's learned by facing hardship and overcoming it.

There are two versions of the theory, Jonathan Haidt, a psychologist at the Stern School of Business at New York University and author of *The Happiness Hypothesis*, told me recently: the "mild version" and the "extreme version."

The mild version: suffering often leads to growth.

The extreme version: we *must* suffer to reach the pinnacle of human flourishing.

"The mild version is well supported by research, while the extreme version is supported by anecdote," Haidt said. Either way, "I believe there is a sensitive period for growth—late teens through early thirties—that if you don't have any major setbacks or adversity during that window, you'll be weaker later on. I think you will be better off with setbacks." Olga's late twenties and thirties were some of the most trying years of her life.

The Toughest Generation

Almost everyone as old as Olga has been stress-tested. They are acquainted with hardship and change of a sort baby boomers can scarcely imagine, from hunger to war to the massive social upheaval of postwar life. They are a cohort that, as Cornell University gerontologist Karl Pillemer, author of *30 Lessons for Living*, puts it, "was pushed to their ultimate limits, as a group."

I have marveled at the toughness of Orville Rogers, a runner from Dallas who is slightly older than Olga. You can see the discipline in the way he moves. He takes long, loping strides, bent forward as if piercing a headwind, breaking his trance only now and again to look at his hand where he's written his target lap times in ballpoint pen.

A pilot stationed in Texas during WWII and the Korean War, Rogers bore, in that second conflict, the weight of a terrible responsibility. An atomic bomb was standing by. Rogers had been told that he would be the pilot to deliver it.

At the world indoor championships in Kamloops I had lunch one day with Sumi Onodera-Leonard, a tiny woman with a cute moon-shaped face and graying pixie-cut hair. She had just won the sixty-meter dash in the Women's 80–85 category, on a bum knee. She needed a proper diagnosis of what was wrong, or at least an injection to relieve the pain. Both were held up because her doctor had just died.

Sumi's toughening up began in her teens, when she and her Japanese family were shipped to an internment camp in Arkansas. Four years later the Second World War ended, and the family was allowed to return home to Southern California.

But discrimination dogged her for many more years. She turned it into fuel. "What you do is, you want to get educated, you want to achieve," she told me. "Because you want to be better than them."

And then there is Ugo Sansonetti, who when we met in Kamloops was the world-record holder in the 60-meter, 100-meter, 200-meter, and 400-meter for men 90 and up. (He's the one who starred in the commercial.)

As with so many others of Ugo's generation, the life he had planned foreclosed without notice. The day after Ugo received his business degree, in the southern Italian hill town of Mottola, Mussolini declared war on France and England, and Ugo was conscripted as a cavalry officer. He and his two brothers, Vito and Giulio Cesare, ended up in exile in South America, and there he became swept up in a two-part dream of his father, Luigi: to build an "experimental human community" and establish a homeland for Italian war veterans. They cleared almost twenty-five thousand acres of rough rain forest in the high mountains of Costa Rica, near the Panama border. They built a sawmill, a landing strip, homes, and cooperative coffee farms. The community grew to become one of the bigger cities in Costa Rica, San Vito de Java, population two hundred thousand.

Ugo has a deep appreciation for how tough and capable his generation is—and how shamefully underestimated it is in a culture that increasingly overvalues youth. Which is why, in 1999, he and his friend and fellow runner Vittorio Colo contrived a demonstration to set the record straight.

Italy's minister for sport had just made a startling declaration. He could no longer condone competitive races by athletes over age 70. Such events were mere freak shows, the minister maintained. The only appeal of watching funny little doddering old folk run was morbid curiosity—maybe someone would blow a hamstring or wander off the track.

Ugo was out of the jungle by then and living in Rome, where he kept himself in top shape with a calisthenics and running regimen inspired by the Italian military. Furious, he issued a challenge to the minister. If the team of octogenarians he assembled ran the 4 × 100 in under a minute, would that bury the issue once and for all? An estimated TV audience of fifteen million tuned in to watch the quartet do it. The idea of banning masters athletics was never credibly raised again—in Italy or anywhere else.

Hardship builds character: so goes the cliché. And if it's true that Olga's generation faced more hardship than generations that followed, their reward was a kind of resilience of the spirit. But is it also true that stress builds resilience in the body? That it toughens us up *physiologically*?

The answer is yes, under the right conditions.

Stress, because it affects us so comprehensively, emerges as a huge part of the picture of aging. How much we've been exposed to, and what kind, and how we innately react to it, and what strategies we've cooked up to manage it better—all these things contribute to the person we ultimately become: strong and youthful or compromised and decrepit.

Avoiding the Permanent Midnight of the Soul

If you stress a rat periodically and in just the right way, you can make it hardier and healthier and even make it live longer.

Put it on a severe diet and the rat may live 20 percent longer than other rats—the calorie restriction seems to trigger survival mechanisms that slow metabolism and actually retard aging.

Hit a young rat with "thermal stress" by plunging it periodically in cold water, and the rat's body will learn to quickly mount a stress response—drowsy repose to DEFCON 1 in seconds—and then just as quickly rebound to baseline once warmth returns. Its vagus nerve is being tuned. A rat that has been thermally stressed as a pup learns not to sweat the small stuff, and doesn't even sweat the big stuff for long. It is folk wisdom that kids growing up on the Canadian prairie are "toughened up" by windchill that freezes exposed flesh in thirty seconds, and fledgling research in humans suggests that what happens to rats may well happen to us. University of Nebraska psychologist Richard Dienstbier proposes that "people who have developed cold tolerance may also have increased their emotional stability."

I've long believed—and await the day someone backs it with hard numbers—in a Saskatchewan Effect. People from that bracing province are overrepresented, at the top levels, in high-pressure creative industries. Troll newsrooms and art departments in Toronto or New York, or writers' rooms in Los Angeles, and you're likely to find, at the still center of the neurotic chaos, a cool, unflappable guy or woman from Saskatchewan.

When I corresponded with Dienstbier, he was just finishing up a book on this elusive human trait of "toughness." He defined it, chiefly, as "physical and mental endurance." Think of things that are hard to do: the ability to keep doing them, to just keep your nose to the stone and grind out the program—that's toughness.

Do the tough live longer? Dienstbier's answer is a qualified yes. Toughness, as he defines it, "strengthens the immune system, with implications for physical health." But there's no doubt that being tougher allows people to live *better*—more relaxed in their own skin.

Young rats don't like to be picked up—it stresses them out. But they get used to being handled, and physiological adaptation follows. Rats handled as pups, according to one study, show "a more muted stress response as adults." Perhaps being the middle child of eleven kids is a little like being a rat that was handled a lot. All that figurative jostling for a teat, that scrabbling for attention and routine interruptions of your privacy and peace, is a hardship in the early going but a boon later.

Acute stress is ultimately good for us, without doubt. If you were designing a life from scratch you'd want to salt in stressors. Speak in front of crowds, interview for challenging jobs, go hungry now and again. Then return to the safe port of routine. Such brief trials, according to Harvard psychiatrist John Ratey, build up "waste-disposing enzymes, neuroprotective factors and proteins that prevent the naturally programmed death of cells."

Chronic stress is different. It is a certifiable life shortener. It has been implicated in cardiovascular disease, impaired immune function, high blood pressure, inhibited DNA repair, and even elevated risk of dementia. (Plus, the things we do to relieve chronic stress—sleeping pills or cigarettes or booze—tend to compound the insult.)

People under chronic stress can become almost incapacitated with anxiety. When the alarm bells never shut off, "the amygdala starts labeling *everything* as an emergency," notes Ratey in his book *Spark*. And now you have a permanent midnight of the soul, with ever-jangled nerves keeping the adrenal glands pumping cortisol, blood-glucose levels artificially high, and inflammation response stuck at "on." The body under stress makes energy available to the muscles, and so shuts down all caretaking functions. It's like draining your bank account to pay a ransom. You'd do it: *once*. If you're asked to do it again and again, there is trouble.

In a perfect world, we'd all want the good stress and not the bad, the toughening-up acute kind but not the breaking-down chronic kind. Olga has had her share of both. But she has largely escaped the kind of chronic stress that defines the daily routine of so many modern urbanites: escalating money

woes and job insecurity and workplace politics we have no control over.   .

Control turns out to be a huge part of the story of chronic
stress.

An ongoing fifty-year study of British government workers
found that lower-ranked employees have mortality rates three
times as high as the managers on top. This seems strange.
You'd think the managers, burdened with more responsibility,
would have more stress, not less. The explanation, some psychologists think, is that it's just less stressful to be calling the
shots. It's better to boss than to be bossed. The dominant
baboon enjoys untroubled sleep.

Literally. Robert Sapolsky, the esteemed Stanford neuroen-
docrinologist, has spent part of every year for almost three
decades observing wild baboons in Kenya, tracking which ones
are the most stressed out and why.

Status counts for a lot. The dominant males, sitting com-
fortably atop a stable hierarchy, are the least stressed—
presumably because they don't have to worry that their dinner
or their mate might be snatched away the moment they let their
guard down. They are masters of their domain, bosses with the
perks but not the responsibilities: a sweet deal. These baboons
are psychologically, and therefore physiologically, healthy, with
low levels of circulating stress hormones and strong immune
systems.

How scalable are these observations to humans? Well,
baboons aren't a bad proxy for us. Not only is their biochemistry similar, but they also have relatively cushy lives and big
brains to use their ample spare time creating problems for
themselves. "Baboons are like westernized humans," Sapolsky
says. "They're not getting stressed by predators but, instead, by
psychosocial hassles from their own species."

For the past half century, Olga has enjoyed something like

that kind of autonomy. She has rarely been "bossed." She has largely avoided the bureaucratic purgatory, the red tape, the disconnect between what you do and evidence that it ultimately mattered (or didn't). The last half of her teaching career particularly was high reward and low frustration. She visited students convalescing in hospitals and tutored them one-on-one. There were none of the riot-squad issues of regular teaching, and few administrative duties.

She lives in a house full of her own paintings of flowers she grew in her own garden. She leaves that house in the morning to pursue a rewarding avocation she chose. The distance between input and output is short. She sees tangible and fairly immediate benefits. Socially she's in control, too. She joins the activities of others, but on her own terms. A meal with her daughter and son-in-law is as easy as a trip up the stairs, if she chooses. She won't be guilted into things she has no interest in. (She doesn't do guilt.) It's about as close to no-strings freedom as you can get without being a seagull or a libertarian.

It's not always obvious how much pressure someone is under, day in and day out. Many people hide their trials well, at least superficially. But the body logs that information. And it will give up a number, if you know where to look.

How Old Is Olga, Really?

The very long-lived find that their age is trotted out as the most noteworthy thing about them, but at the same time the number slowly becomes meaningless. They have entered a statistical black hole. Is this drug beneficial to them? (Who knows? Pharmaceutical companies typically avoid clinical trials on people born before the Depression.) How high can they safely rev their heart? (Who knows? Physicians typically play it safe with recommendations for what senior citizens ought to attempt.)

Can they get travel insurance that's not astronomically expensive? (No.)

Now think how the answer to these questions would change if you could prove that while your birth certificate says you're 90, in body you are effectively only 65. Biological, rather than chronological, age obviously matters most. But how do we measure it?

In India and Southeast Asia, physical flexibility is often considered a more accurate metric of age than years. (By which reckoning Olga is 50 and I am 113.) Some researchers tally up health deficits and plug the results into a complicated "Frailty Index" that Canadian geriatrician Kenneth Rockwood and his colleagues developed. The geneticist Angela Brooks-Wilson boils it down more simply. Your real age, she says, is "what you can still do." That functional definition feels right, but it's imprecise.

The best biological clock we know of requires very high magnification to read. It exists inside our cells, at the end of our chromosomes, where little protein caps seal the ends, protecting the DNA.

Those caps are called telomeres. Every time a cell divides a sliver of telomere is sloughed off, until—some fifty to seventy-five cell divisions later—the telomere is too short to do its job. The chromosome is exposed. The next cell division starts cutting into the principal. Genetic damage results, leading to cell death, which we see as age-related disease.

Everyone's telomere length is set at birth, a gift of inheritance for better or worse. But life circumstances thereafter can change the burn rate—a lot.

In a 2011 study, researchers at the University of California at San Francisco interviewed parents who were raising autistic kids—a stressful proposition. Blood samples of those beleaguered parents were then sent across campus for analysis, to the lab of Elizabeth Blackburn, a cell biologist who won the Nobel Prize in Physiology or Medicine in 2010. Those parents' telomeres were short; their cells looked ten years older, on average, than the cells of age-matched parents of "neurotypical" kids.

Very short telomeres probably signal one of two things, neither of them good. Either a genetic defect is present or your body is so stressed by some fight (with a person or a circumstance or a disease or an infection) that it's aging you prematurely. Either way, your cells have been dividing way too rapidly, and without some intervention you're looking at an early death. "Telomere length," says Gary de Jong of the telomere-measurement lab Repeat Diagnostics, "tells us something about what's going on in our body and in our life."

Not long ago a FedEx package arrived at that company's office in North Vancouver. In it were two blood samples: one from Oprah Winfrey and the other from her colleague the physician Mehmet Oz. Oprah was putting together a show about aging.

In the episode, Dr. Oz held up a chart. There was the curve of "normal" lymphocyte telomere length plotted across an average human life span. A big dot sat a bit above the curve: Oprah's score. She has somewhat longer telomeres than usual for her age. To estimate her "biological" age, Dr. Oz drew a straight line from the dot back to the line. Oprah was in good shape: by this metric, she was in her late forties, not her late fifties. (Of course, we don't know how long Oprah's telomeres were to begin with. What you'd want to do, for a more accurate snapshot of the *rate* she's aging, is repeat the test at intervals.)

If Oprah had long telomeres—a partial explanation, perhaps, for her youthful pep—Olga must have *really* long telomeres for her age. The longer the telomeres, the more robust and stress-resistant their owner, right?

To test this theory, we sent a vial of Olga's blood to Repeat, and then sat back and waited. It was a pregnant two weeks. Something about the process felt vaguely illicit—like filing an Access to Information request with God.

I was as curious about my own cellular age as Olga was about hers, and had planned to get my telomeres measured, too. But the test requires a signed release from a doctor, and when I visited Rocky, my family GP, to ask for it, he provided some sober second thought.

"Be careful about turning over stones," he said. What if my telomeres are really short? That information is now on my permanent medical record. It could flag me as a health risk, which could matter to a future employer. I might not be able to get life insurance. Our impulse to investigate our physical health in ever-finer grain—to drill deeply into our genome with the evolving technology that allows us to do these things—has implications we may not have thoroughly considered. Knowledge has both benefits and potential costs. In the case of the telomere test, it was worth it for Olga to proceed, but not for me.

When the lab got in touch with the test results, they were surprising. Olga's telomeres are about average for her age. Just a shade longer. If you drew a line back from Olga's dot to the curve, as Dr. Oz did for Oprah, you'd peg her biological age at around 78. Still good, but not the extraordinary discount I had expected.

Angie Brooks-Wilson, the geneticist and cancer researcher, wasn't surprised at the number, though. She has seen this movie before.

When she tested her inordinately healthy and long-lived Super Seniors for telomere length, they mostly scored about the same as Olga—not short but not too long, either. And unlike the scores of middle-aged controls, which were all over the map, the Super Seniors' scores were tightly bunched, remarkably consistent. "We think they are clustering toward a sweet spot," she says.

It turns out that while short telomeres spell trouble, so, too, do very long ones. "There may be benefits to shutting things down after a certain number of cell divisions," Brooks-Wilson says. "Very long telomeres allow the chromosomes to have more mutations. You might get a tumor."

That is called *the fine print,* if you're a consumer of a class of experimental drugs called "telomerase activators" now flooding the market. Such drugs—propelled by dreams of ageless tomorrows—are said to spur production of the enzyme that protects and restores telomeres, so that in theory they never shorten, and cells can go on replicating forever. It's one of biotechnology's hottest research avenues, a multimillion-dollar business whose kudzu growth obscures a basic question: Can you really hack natural aging through genetic engineering this way? Or does the quest for Endless Life route, awkwardly, through a station called Early Death? Clearly it's a gamble plenty of tech-age survivalists are willing to take.

Few of us—let's face it—share this lot's urgent desire to live for hundreds of years. Fewer still care if we're still race-fit beyond age eighty. But many of us do worry that we're one day going to start putting the dirty laundry in the oven.

More than dying, boomers fear losing their marbles and becoming a burden, many surveys reveal. Whatever ways we can reduce the likelihood of *that*—the close to 50 percent chance we have of being clouded by dementia when we die—will be front-page news. In this respect, Olga's brain is at least as much a source of intrigue as her body.

It's time to take a peek inside it.

3

Tests of Mind

Reprogramming Olga's Brain

AT THE TRACK behind the junior high school near her house, Olga and Barb Vida, the only real coach Olga's ever had, are going through a careful warm-up. After parting amicably eight years ago, they have reunited today for a high-jump lesson.

Side by side, the two move deliberately across the field. Olga, mirroring Vida, high-steps it through fairy rings of mushrooms that have sprouted during the interminable June rains. Vida brings her knees a smidgen higher with each stride, stretching the hip flexors and surrounding tendons. "These tendons are stupid," Vida says. "They only know how to do what they've done before."

Vida is a beautiful 50-something Hungarian firecracker who talks in an unbroken stream of encouragement and directives and facts. An alternate for Hungary's track team in the 1972 Summer Olympics in Munich, she went on to coach, tutoring a young jumper named Mike Mason to a junior world title. But her star pupil has been the world's oldest competitive female high jumper: Olga.

The pair began working together a decade ago. Olga's first

mentor had died, and she felt she needed more instruction than she could get from library books. Olga and Barb were kindred spirits, in a way. "We have kind of similar stories," Vida told me. "I divorced, with two children, and ended up in a different country." Vida left Hungary in 1991 "and burned everything behind me."

Vida put Olga on a routine of staggering intensity for someone then approaching 85 years old: three workouts a week, each two and a half to three hours long. Olga did punishing isometric exercises such as the plank and the roman chair to develop a rock-solid core, a pillar of strength down the middle of the body upon which all movement depends. She did squats, with weight on her shoulders. She rode a stationary bike for cardio. On off days she did lighter workouts alternating "general strength" with "core strength." (Sundays, then as now, were sacrosanct: she rested.) "She got very fit—extremely fit," Vida says. "They actually called her 'Dynamite Girl.'"

And through those years something extraordinary happened. Not only did Olga stem the decline in strength that happens to just about everyone in their 80s, she got better.

It's a coach's job to figure out how to motivate her students. But "that was never something Olga needed," Vida says. "I actually had to try to hold her back." Vida tried to put a governor on Olga's weight-training ambitions. "She would have gained more muscle mass, and if you have more muscle mass, you have to lift it." Olga set her sights on the marks of masters legends such as Swede Nora Wedemo and Aussie Ruth Frith—and when she hit 85, she knocked those records off.

Vida felt Olga was doing too many events. When she heard Olga was thinking about adding a twelfth discipline—pole vault—the two banged heads. Vida said she simply would not allow it. "I also didn't want her to do the hammer but she fell in love with it, so what're you gonna do?" Vida told me recently.

"She's so determined. So goal oriented. I never met anyone like her."

Toward the end of her time with Vida, Olga began to accept the bridle (or bridled herself). She drew back from heavy-intensity training—and the trend continued even after Olga and Vida parted ways. Vida has watched the dialing down of intensity with mixed feelings. On a personal level, "she is even more clear and more relaxed than she was before," Vida says. "She is wise. She's like a fruit that is ripe. I get goose bumps just saying it." But Olga's core strength has diminished, and Vida has noticed a dip in the level of confidence driving Olga's body through space.

That's what today is about. Vida wants to make a technical change that will get her jumping higher. Today is about reprogramming Olga's brain.

Olga has lately been using a "scissors" technique to get over the bar. It looks as it sounds. You approach the bar sideways and sort of sidle over it, one-and-a-two, staying upright. The scissors is a relic of a track-and-field era when pits were sawdust and sand, so landing on your feet was advised. Today, when a superfoamy mat awaits, all top high jumpers use a technique called the Fosbury Flop. That's where—counterintuitively—you turn away from the bar at the last minute, arch your back into a cupid's bow, and sail over backward. It's a trick of physics: the jumper's center of gravity actually passes under the bar.

The thing about the flop is, you've got to be flexible. Eventually, the spine stops being able to arch that way. Nobody over age 75 does the flop.

So here was the problem for Olga. The scissors is a fairly explosive technique; to use it profitably depends on a level of power that Olga no longer has. But if the scissors is inappropriate, and flopping is out of the question, what's left?

Vida's answer is the Western Roll. That's the one where you go over the bar facing it, curling around it like a rotisserie chicken. It's a more graceful motion than the scissors and less dependent on the elastic stretch and release of muscles and tendons in the hips. It depends on timing and rotation—faculties with no shelf life.

Vida reckons Olga could probably add fifteen centimeters to her world record by mastering the Western Roll—thereby establishing a mark so high it would be untouchable.

But here's the thing: Olga already knows the Western Roll. Vida taught it to her years ago, and Olga used it to smash records in her 80s. Then something happened. Olga developed the bad habit of putting a hand down on the mat before her body landed, and when you do that a guy in a white shirt holds up a red flag. Olga was disqualified enough that she grew discouraged and just stopped doing it. And then she forgot how to do it. Hence this refresher.

Vida demos the run-up to the jump she wants, lifting her arms with an exaggerated motion. "I'm trying to do it exactly as she's supposed to do it, so she can copy," she tells me.

Olga takes a couple of bluff charges at the bar. She strides toward it and then pulls up, like a spooked horse at a water jump. High jump is the most "mental" of all track events, in her view. She has to psych herself up for it. To force herself through the impasse, she concentrates on just two things: *lift those arms* and *jump!* She comes in again, but without full conviction, and lands on the bar.

"You can't hesitate," Vida tells Olga. "When you hesitate you use energy."

Olga picks herself up.

"How's it feeling?" I ask.

"Well, nothing hurts," she says. "But I'm not confident."

And it's true; these jumps look dicey. Olga is just barely landing on the edge of the mat, and sometimes sliding off it onto the track.

ALL this is happening in split screen. Because today Olga is sharing the mat and the coach with a local 15-year-old prodigy named Kenny who just won the heptathlon in the provincials. Kenny's best jumps are years ahead of him, but he is still clearing more than 1.8 meters—which would be a world record for men over 55. He's working on his flop.

Barb is teaching Kenny to scorn the bar. "As you clear it look down on the bar and say, 'Pah! You are nothing!'" she tells him. As Kenny comes gliding in on his approach, Barb says, "Make yourself tall!" I noticed she didn't offer that advice to Olga—who it seemed could have used it more, being at least fourteen inches shorter.

Olga and the prodigy alternate jumps. Kenny is loping in from the right, describing a graceful J with his run-up and exploding up and over. Olga is all business from the left. She is thinking about jumping, trying to "get it in the computer," she tells me, touching her temple.

I turned to Vida. "Won't it confuse her body?" I ask. "Since she's now so used to the other technique?"

Vida shook her head no. "The scissors and the Western Roll are very different movements." The body gets confused only when you feed it something that's *just a little different* than it already knows how to do, she explains. The two ideas clash, like discordant notes. Vida is positive that this will work, because the roll is inside of Olga's brain, still. It's there in the snake-shaped Purkinje cells of her cerebellum, where muscle memory is stored. She just needs to call up that map. And as she imagines the movement, and actually performs it with laser attention, she will burn the new movement into updated memory.

Sometimes Olga surprises me. That she would make this adjustment to try to go higher, at this point in her career, seems a minor miracle of will. That it could work—that this new technique could be rewired into her brain—that's a minor miracle of human anatomy.

NEUROPLASTICITY is one of science's most startling discoveries of the past thirty years.

We used to think brain development was a one-way street: you were born with *a thousand trillion* neural connections, give or take, and what followed was mostly a lifetime of pruning, according to the rules of use it or lose it. We now know that the brain can retrofit itself, growing new neural connections (and plumping up existing ones) upon exposure to novel circumstances. That's what adaptation is all about.

True enough, it works best if you're young.

During the high-jump lesson, Vida looked at Kenny the way a land artist looks at mud. Here was a 15-year-old with the body of a crane and the brain of a sponge. Vida made some tweaks to Kenny's technique and within fifteen minutes he was jumping four inches higher. "See what you can do with the young ones?" she told me, sotto voce.

As we age, our brains do get less plastic. At the same time, we develop shortcuts and habits that streamline the process of living, and these do our brains no favors. "Creative-class" boomers flatter ourselves that we are "lifelong learners" thinking outside the box on a regular basis, but we really aren't. "We may deceive ourselves into thinking we're learning as we did before," writes Norman Doidge, a physician and author of *The Brain That Changes Itself.* "But we're not." The things we may do habitually to "stay sharp"—such as reading the paper or indulging a crossword-puzzle habit—are mostly just a replay, he says, of skills mastered long ago. The part of the brain whose job it is to secrete the neurotransmitter that helps us tune in and form sharp memories? Well, that brain region sleepwalks through tasks we're doing by rote. And then, failing to receive any further orders from HQ, that region starts shutting down.

As we get older, plasticity just takes a little more conscious effort. Arthur Kramer knows this. The cognitive psychologist, who's pushing 60, will sometimes eat with his left hand or drive a different way to work or punch in a different music station to throw a little productive confusion into the wetware.

Pumping Gray

Kramer, who goes by "Art," is director of the Beckman Institute for Advanced Science and Technology at the University of Illinois at Champaign-Urbana. For twenty-five years, he's been working a seam where aging, cognition, and exercise meet, and so has become one of North America's leading voices on the relationship between, as he says, "body fitness and mind fitness."

Kramer has more than the usual academic interest in Olga. The son of a professional light-heavyweight boxer, he himself was an accomplished multisport athlete in high school (a 4:30 miler and a pole vaulter) before his knees blew out. If he sits for any length of time, those surgically repaired joints greet him with pain when he stands up.

He has spent much of his career trying to bury the stereotype of the "dumb jock." As far back as Ancient Greece, the widespread thinking was that testing the cognition of someone like Olga would be worthless, because there wouldn't be much of a brain in that head. "All natural blessings are either mental or physical," the ancient Greek physician Galen once mused. You got one at the expense of the other. Not everyone believed it. The philosopher Plato—himself an accomplished wrestler—

was sure that bodily fitness and mental fitness worked together. The goal, indeed, was to bring those two realms "into tune." Kramer is like Plato, if the philosopher had had access to an MRI machine and a well-staffed lab. In 2006, that lab nailed the circumstantial case between exercise and improved cognition.

Kramer took 60- to 80-year-old test subjects who were thoroughgoing couch potatoes—"these were people who didn't even walk to stores," Kramer says—and put them on an exercise program, starting with a modest fifteen minutes of walking per day and slowly increasing it to forty-five minutes. After six months, their brains had grown. Substantially. And not just in the hippocampus, where memories are consolidated and new neurons are known to grow, but in the frontal and temporal lobes as well, where reasoning and sensory processing take place. There was more gray matter (the neurons of the cortex) and more white matter (the connecting pathways). The exercise stimulated all aspects of cognition. It fertilized the whole farm. The subjects showed improved reasoning, spatial function, processing speed, learning, balance, and several kinds of memory. Chatter between brain hemispheres increased. Decision making improved 15 to 20 percent.

On a stifling day in July, we cruised into the air-conditioned confines of Kramer's office at the Beckman Institute. He seemed pleased to learn that Olga doesn't come from a long line of Methuselahs. "We know it's not just your genes," he said. Since body and mind amble lockstep into old age—things that insult the brain insult the body, and things that benefit the body generally benefit the brain—her lifestyle choices must somehow have left their mark in those gray coils.

A battery of diagnostic tests lay ahead.

"We want to look at your cognitive skills and your ability to think quickly and correctly—in terms of attention and memory," explained research scientist Dr. Laura Chaddock, who had joined us in Kramer's office. Chaddock, a freckled former tennis prodigy who has become a dynamo of productivity at Beckman ("If there is an enthusiasm gene, she got it," says Kramer), designed a number of the studies. Tests would reveal Olga's "fluid intelligence" (thinking on the fly), her "interference control" (focus on one thing while being distracted by another), and her processing speed. All of these faculties are known to decline, in a curve that steepens with depressing predictability as we age.

Brains—like pretty much everything else in our bodies—shrink over time. It starts happening around age 20, when cells start dying faster than they can be replaced, and connections between neurons wink out. The average 90-year-old has around forty-seven thousand miles of "wiring"—the signal-sending axons of neurons in the brain. That sounds like a lot, but it's only half of what he or she once had.

That wiring degrades, too. The axons of the neurons lose their protective insulation, so signals are sent more slowly and communication between neurons isn't as tight. It's like being downgraded, by your Internet provider, from broad band to dial-up.

Names and phone numbers migrate from the tip of the tongue to the dead-letter office. And this is quite apart from the big, scary cognitive afflictions that become much more prevalent beyond age 80.

"You've dodged more than a few bullets," Kramer told Olga in his office earlier. "You don't have Alzheimer's—I can already tell that having talked to you for fifteen minutes." He turned to me. "Now, Bruce and I, one of us probably will get it," he said, more jauntily than seemed strictly necessary. "*If* we live long enough."

Olga would be by far the oldest subject ever to go into the Beckman magnet (the MRI machine), which posed a tricky methodological problem. There are really no other apples to compare her to. "The brain wiring of a 93-year-old world-class athlete is unknown in the neuroscience literature," says Chaddock. What Kramer and his colleagues learned about Olga would be the beginning of a grander, international project as other Super Seniors join the global database, and scientists deepen their understanding of optimal human aging.

Olga's Crossing

On the outskirts of town, planted amid the cornfields, sits a facility called the Illinois Simulator Lab. Here Olga's ability to multitask would be ingeniously tested.

Inside is a virtual-reality chamber known as "the Cave." In the center of the room is a treadmill. In front of the treadmill a huge projection screen rises and wraps around on each side, so that when you're on the treadmill you feel surrounded. Add 3-D goggles and you are convincingly transported to another world the moment the lights go out.

Olga donned the gear and was suddenly looking out upon Springfield Avenue, a busy two-way street near the University of Illinois campus.

It was rush hour. The street teemed with motorists who apparently missed the lesson in driving school about stopping for pedestrians at crosswalks.

The tester, Aubrey Lutz, a chatty young redhead with a unicorn tattoo and high heels so sharp they punched perfect circles in the foam safety mat when she accidentally walked across it, put a hand on Olga's shoulder. "So the goal," she explained cheerily, "is to try to cross the street without getting hit."

Cars emerged from both directions at unpredictable intervals. What contributes to the spooky realness of the test is that the treadmill drives the virtual-reality environment. When you just stand there, the only thing in motion is the cars. As you walk, the streetscape adjusts; the faster you walk, the faster you move through that space. The better you get at crossing the street safely, the faster the cars start to come. The test adjusts to your success.

Olga stood on the treadmill and surveyed the scene. Springfield Avenue loomed like a wide river.

To a lot of people, this simulation is like a video game—a fantastically realistic one, and yet a game just the same. But Olga doesn't know from video games. She has never played one. So unlike just about everyone else who has stepped into the Cave, she felt no safety in the artificiality of the setting. It was real.

"Olga, I'll tell you," Aubrey said. "Everybody gets hit by cars in this test. It's part of the learning process. It just makes you a stronger person." She laughed. But Olga didn't. "If you get hit by a car you will hear a crash sound and the screen will go red. There's no blood or gore. There's a little tragedy in the color."

Olga stood on the curb. Looking left. Right. Left. Eventually the electronic voice of the machine—it sounded exactly like Stanley Kubrick's "HAL"—shattered the silence. "You're too slow." Red screen. Game over.

Trial two. Olga's face was tight with conflicting emotions. She was flushed. She was alive to HAL's impatience. She stepped out—into too small a gap. Crash! "You've been hit," HAL said. "Better luck next time."

"He was going far too fast!" Olga said angrily.

Trial three. Olga was in a bind. She wanted to cooperate with Aubrey and HAL and just *do* this thing. But that would mean somehow overriding everything her senses were telling her. And her self-preservation instinct was so strong that she simply could not step off that curb. She shook her head at Aubrey.

"I don't see myself venturing out under these conditions," she said, evenly. "There's not enough time."

A bit of debate ensued among the testers. If Olga's intensely rational brain wouldn't let her past this first stage of the experiment, then the test couldn't proceed to the second and crucial stage where she's saddled with distractions. A technician was summoned to tweak the software program and slow down the cars.

While he was working, I nipped next door to a room tricked out with a flight simulator. It has been used to test, among other things, whether age is a factor in the performance of commercial airline pilots. Should passengers prefer a young pilot, whose myelinated axons inside his wrinkle-free noggin produce blazingly fast reaction times? Or can an older pilot's experience compensate for the loss of youth and quickness? (Turns out it's a draw. "Experience," the study of those pilots' performance concluded, "can level the playing field.")

A technician gestured to the empty cockpit chair. A program was loaded and ready to go. I was piloting a Beechcraft King Air B200 twin prop plane. The challenge was to land it on Meigs Field, a now-defunct airport tacked picturesquely, precariously, onto the Chicago waterfront. I got a little instruction and then suddenly was airborne. Passing over the city, punching through thin clouds, splitting the uprights of the Hancock building and the Willis Tower, I banked out over the lake, took a deep breath, and came in for a landing. Steady. Steady. On target more or less. But at the last minute, I dived too steeply and crashed—and the moment was so gut-wrenching I almost fainted. For a split second, in these circumstances, you forget it's a simulation, no matter how firmly you *know* it is. It's intense. And I was raised on video games. I could only imagine how terrifying the street-crossing scenario was for Olga next door.

Her machine was up and running again and she was actually doing better when I checked in. She looked left and right and left, very quickly, like someone watching competitive Ping-Pong, and then she committed. She didn't walk briskly across that road; she went for it, full-out, like a squirrel—the treadmill whining from the high rpms—and she didn't let up till she was safely on the far sidewalk.

"Good one!" Aubrey said.

Soon Olga had progressed to the "distraction" scenario. Aubrey fitted her with headphones. Now she'd cross the street and take a pretend cell phone call at the same time.

This condition compounded the ridiculousness. Olga doesn't own a cell phone. And even if she did, she said, she would never talk on it while crossing a street. Do people do that? That's nuts.

The first time Olga tried crossing while on the telephone, she was hit. She was just not sure she had the RAM to do this safely. Now she was at the curb again. Looking left, right, left, right.

"So, Olga, what was Saskatchewan like, where you grew up?" Aubrey asked.

"Flat, like here," Olga said, narrowly avoiding being greased by a red sports car. "Look, I can't talk to you, dear—I've gotta concentrate."

The questioning by the chatterbox tester continued unbroken, and Olga's answers started to become sharp: "Okay, Aubrey, I've gotta watch here!"

Eventually Olga's decorum broke down. She said: "Are you trying to get me killed?!"

In the end, Olga ran forty trials. With the way she was approaching this task—as if street crossing is to be a demonstration sport at the Rio Olympics—the cumulative effect was like having run forty races in a row. Sweat rilled down her face.

As Olga stepped off the treadmill, Aubrey asked her her age. She hesitated for at least eight seconds. "Ninety-three and five months," she finally replied. Then she sat at the back of the room and did deep-breathing exercises to settle her jangled nerves.

HALF an hour later, in an air-starved little interview room back at the Beckman Institute, Olga sat down and opened her brain for business again.

A young blond graduate student named Andrew Lewis prepared a comprehensive cognition test called the Salthouse battery. Here Olga's memory, IQ, and spatial reasoning, among other faculties, would be appraised. She was red-eyed. She hadn't slept much because she had stayed up too late doing Sudokus.

Lewis administered a memory test that was essentially him telling her a very short story and Olga repeating the story back to him in as much detail as she could muster.

Some details she retained and some she forgot. She couldn't remember, for example, that it was "Anna Thompson" of the "South Bronx" who was robbed in this story, or precisely the hour of the day when it happened, but she remembered that the local police were touched by Ms. Thompson's plight and took up a collection for her—an emotionally resonant detail.

When Olga made errors they almost invariably improved the story, punching up the characters or hinting at their motivations. Her memory is a sympathetic editor.

She did less well on a test of randomly paired words, where there were no semantic links to lean on. Those simply didn't stick, and Olga pronounced it "crazy" that such words would be combined for no reason. Aging brains can inductively reason their way through the fog, but such brute memory tasks are the province of the young.

What *I* won't forget is the dynamic between Olga and her examiner. Lewis had carefully explained the instructions, and then sat back and listened. This was premium-quality listening, backed by nods and eye contact. The full attention of this almost unlawfully handsome young man seemed to be fueling Olga somehow, helping her to hang in there.

"I'd like to take that one home with me," Olga quietly joked afterward as she gathered her things.

In the basement of the institute, Olga emerged from the change room in a blue scrub gown. She'd been asked to remove all rings, earrings, and jewelry, which could interfere with the electromagnetism about to flood through her body.

"Brain scan time," someone said.

"I hope they find one," Olga said under her breath, as she entered the chamber. The door closed behind her.

Olga was swaddled in a warm blanket, fitted with headphones and a broadcast microphone, and slow-motored into the maw of the machine. She was pretty relaxed. Usually, if you're about to get a brain scan, it's because something is very wrong. It's a privileged position to be getting one because everything is apparently so very right.

Inside the MRI machine Olga was distracted by a variety of strange sounds: a low chug-chug-chug, followed by assorted whirs and bangs. It was hard for her to make out the directions arriving through her headphones.

On the research side of the glass windows, grad students and technicians with toned calves and all-terrain sandals attended to computer monitors. Blooming now on-screen was a kind of bird's-eye view of Olga's brain—three images from three perspectives.

"Wow," said Gillian Cooke, a postdoc research associate. "I can't believe it.

"In the brain of an older person there's usually lots of atrophy along the sides," Cooke said. In old age, the gray matter typically shrinks back on itself, away from the skull, leaving a noticeable gap there. "Here I don't see it."

"Her white matter tracts"—which carry information throughout the brain—"look well preserved," said Chaddock. "I bet her functional brain networks are well connected, too.

"Ventricles"—cavities in the center of the brain filled with cushioning cerebrospinal fluid—"usually grow dramatically," Chaddock added. "But hers look fantastic."

"We're currently working with middle-aged adults," said Cooke, "and a lot of the brains we're seeing aren't as well preserved as this."

Andrew Lewis, who had been silent till now, piped up. "I'd better start running more," he said.

Among all the people who get MRI scans, older athletes tend to be among the worst subjects. You have to be absolutely still in there. But it's torturous to be still if you've hurt things, so "older athletes tend to wiggle around," said Kramer, and that muddies the image.

Olga was quite still. Even after forty minutes she was still. Was she okay in there? "Olga?" said Laura Chaddock. There was no response. A little louder: "Olga?" Nothing.

She had fallen asleep.

The Secrets of the Sisters of Notre Dame

The Beckman researchers were coming to believe they were dealing with a fairly remarkable brain here. If so, the question is, *Why* is it remarkable?

Exercise is obviously part of the story: it's the reason Olga is a little bit famous, the reason we're here at all. But it can't be the whole story. If it were, the upper age brackets of masters track meets would be little micropopulations of spry old folks entirely, magically protected from dementia. They aren't.

At the World Masters Indoor Athletics Championships in Kamloops, British Columbia, I watched a 90-year-old gentleman, shivering in the chilly morning air, prepare to throw the discus. He took a compact little backswing and tossed. It landed 10 meters away. He started slowly walking toward it—which would have disqualified him—until he heard his daughter on the sidelines hollering "Out the back, Dad!" He stopped and smiled and gave the gotcha signal and backed out and around. ("Sometimes he forgets," she whispered to me.) Last year, in the middle of a race for men 80 and up, one of the competitors suddenly remembered that he hadn't registered for his next event, the shot put. And so as he came around the curve he tangented right off the track toward the officials' tent to do so. A couple of minutes later, having filled out his form, he jogged back onto the track to finish the race, which was still in progress.

Something about Olga seems to be offering her protection from the ruthless statistical likelihood of cognitive impairment. Something or some *things.*

In St. Paul, Minnesota, there is a group of Catholic nuns who have grabbed headlines for being not just extraordinarily long-lived but extraordinarily spry and sharp. Many of the School Sisters of Notre Dame have been centenarians—a group that ought to have a coin-toss chance of serious cognitive impairment. Yet in this particular convent, evidence of dementia is almost entirely absent.

As the sisters have died their brains have been examined, with often shocking results. Many look like an old house full of wiring nowhere near up to code. One 88-year-old nun had a brain so full of myeloid plaques you would have guessed she had full-blown Alzheimer's. Two possibilities emerge: either plaques aren't the culprit in Alzheimer's after all, or else plasticity is nature's most ingenious contractor. Those apparently ravaged brains were webbed with workarounds. The new wiring actually takes up a lot of real estate in the brain. It makes the brain bigger, swollen with new white matter—the connections to the gray matter, where processing takes place.

These brains become like nimble little companies that keep having to stretch and improvise in the face of ongoing layoffs and funding cuts. The size of a brain thus hints at the cunning it took to grow it to those dimensions. It is a kind of proxy measurement, suggests Harvard's John Ratey, for something otherwise impossible to measure: wisdom.

The sisters, predictably, have been of intense interest to scientists and social scientists. Even if those plaques turn out to be a red herring, the nuns are obviously doing something right to maintain their brilliant mental functioning. But what?

As a group, the sisters have a lot in common with Olga.

They tackle crossword puzzles and other brainteasers daily. Olga does Sudoku daily. Like other logic puzzles, Sudoku sharpens cognition in certain ways. You have to hold things in memory while processing other things, and recognize patterns from partial clues, and as you get good you can keep cranking up the challenge level to continually surf the edge of just-manageable difficulty. (Which may be optimal for brain development. It's certainly optimal for motivation. Studies show our interest in a task peaks when the risk of failing at it is about fifty-fifty.)

The sisters are almost all college educated, and many have been teachers. Olga, too. All those nights spent completing her bachelor of education—becoming the only one of the eleven siblings to get a university degree—may have paid off for Olga beyond just the sense of accomplishment. There's evidence that time spent in the classroom actually correlates with better brainpower. For whatever reason (a prevailing theory is that postsecondary education stokes curiosity and promotes habits of lifelong learning), people with postsecondary education actually have denser brains than those without it. Brain density has been linked to a smaller decline in cognitive function as we age—probably because there's just more "bench strength" in dense brains, more neural connections to back up the ones that foul out.

Carl Cotman, a neurologist at the University of California at Irvine and director of the Institute for Brain Aging and Dementia, Research and Graduate Studies there, found education levels to be one of the common denominators in older subjects who forestalled cognitive decline. (Another was attitude—about which more in a few moments.)

The nuns and Olga both have solid lifelong diet and sleep habits—two things often missing from the regimes of older people. Poor sleep and lousy diets wreak havoc on the brain, sometimes permanently. Sleep deprivation guts working memory and shortens life span. (How much? One long-term study of fifteen thousand American nurses, published in 2012, put a number on it: two years.) As for diet, a regime deficient in nutrients and healthy fats can increase the chance of the small strokes that impair brain function.

Olga shares some other lifestyle habits with the sisters: she doesn't smoke, or overeat, or let herself get chronically stressed out. And she has a few going for her that the nuns don't. For instance, she gets out more.

Even before she began the peripatetic life of an athlete defending her world titles, Olga had been around the globe—by taking advantage of a teacher swap that took her overseas. Travel is a barrage of multisensory stimulation—new languages, tastes, smells, and customs, as well as orienteering challenges. "Enriched" environments like that, many studies have shown, promote plasticity. And navigating new territory, or even contriving new routes in old territory, has been shown to grow the hippocampus—which plays a big role in memory.

Olga is bilingual. Learning languages—like anything requiring focused attention on something novel—makes brains plastic; it causes new neural connections to grow like asparagus. Knowing multiple languages "also protects you [from dementia] as you get older," says Kramer. "And having multiple languages strengthens cognition just generally—even for skills unrelated to languages." There's some evidence that the earlier that languages are picked up, the stronger the protective effect in old age. It's wild to think that the Ukrainian she learned ninety years ago could be helping buck up Olga's brain now, but it sure looks that way.

The sisters get regular daily exercise; Olga gets more.

We don't know exactly how those various brain-building attributes work in combination—just that it appears they do. When you feed the brain in more than one way, multiplier effects emerge.

One recent study, out of the Mayo Clinic in Rochester, Minnesota, had seniors—aged 70 and 93—doing brainteaser-like puzzles on a computer. These did indeed seem to give the old folks a small cognitive leg up, but the small effect became a big effect when physical exercise was added to the mix. If they danced or played tennis on the side, subjects got a kind of super-boost from the brain games. Somehow the two processes appear to work in concert. Another recent, well-designed German study of subjects 75 and older arrived at a similar conclusion. One group was assigned cognitive training, and one exercise, and a third both. On IQ tests, the group that did puzzles *and* exercise blew the other two out of the water. (Turns out Plato's hunch was right: good things happen when mind and body come "into tune.")

One of the great mysteries of neuroscience is why the same

neurodegenerative disease affects different people so differently. Hold everything else constant—age, gender, lifestyle, genetic vulnerability—and we still see huge discrepancies in the impact of devastating diseases such as Alzheimer's. Something is offering the high functioners assistance.

That something is what Columbia University researcher Yaakov Stern has dubbed "cognitive reserve." It must be the Holy Grail of every aging boomer. Cognitive reserve is both a physical thing and a set of strategies: at the neurological level, more brain tissue and more connections, and at the strategic level a sort of learned wiliness. "You're now clever and experienced so you can think of strategies to get around the limitations," says Kramer.

"We build up cognitive reserve over a lifetime," says Kramer, "because we're physically fit. Or we wind up getting a lot of education. Or we eat right. Or we have big families and lots of social interaction. Or some combination of all of that. And then we can use that reserve and pay it out when bad things happen." Think of cognitive reserve as a kind of currency, like psychological retirement income. The more you have, the better you'll be able to weather bad fortune, such as Alzheimer's disease.

The work of Kramer and others leads us gently back to one thing. Exercise can't be the whole story, as noted. But a strong case can be made that, among all the factors we have considered, exercise is the one driving the bus. Where they've been measured head-to-head, nothing can match it.

Ladies and gentlemen, start your engines.

4

The Sweat Prescription

The Best Brain Builder of All

IN A GROUNDBREAKING 1999 study published in *Nature*—and then in subsequent studies over the next dozen years that built on it—Kramer and his colleagues proved that exercise is heavy artillery against age-related cognitive decline. It does what dry heat does to the mold in your attic. It halts the damage and starts clearing things up. "We're not just talking about stemming cognitive decline, here," Kramer says, "but actually improving cognition."

Recently, Kramer's colleague the experimental psychologist Justin Rhodes tested exercise against "environmental enrichment"—widely believed to spur brain growth, and a darling of many educators and progressive environmental planners. Rhodes put rats in an enriched environment with plenty of novel toys and features, so that the rats were "hearing and smelling and playing with new things all day." Another group exercised in their drab old digs.

The result? Both groups increased their brain plasticity. But only one group grew their hippocampus: the exercisers. "In the enrichment group, we didn't find any more neurogenesis than in the sedentary controls," Rhodes says. "On that one variable—growing the hippocampus—exercise mattered and novelty didn't."

Even the vaunted London cabbie effect—where sharpening your brain map of an area by navigating new corners of it all day has been shown to grow the hippocampus—can't compare to exercise. In terms of growing the hippocampus, exercise, says Rhodes, "would blow it away."

Strong claims are made for the brain-building effects of Sudoku-like brainteasers. Michael Merzenich, an emeritus professor of neuroscience at the University of California at San Francisco, touts his own neural exercise program called Posit Science for helping his typical 80-year-old subject act, "operationally, more like someone of 50 or 60." Brain training is a multibillion-dollar business, and boomers are on it like castaways on coconuts. But not everyone finds the data convincing. Psychiatrist George Vaillant, director of the Study of Adult Development, which follows a group of Harvard alumni over their lifetimes, points out that the studies linking puzzles to longevity are mostly short-term. "When you have the full record," he told *Maclean's* magazine recently, "you find that people who are getting Alzheimer's stop doing puzzles." Moreover, a recent meta-analysis by Norwegian researchers found that the gains people make on such "working memory" brain games don't necessarily carry over into real life.

"You can definitely learn new things," says Rhodes, "but they're not going to help you as you age, with things like, you know, driving. These are real things that people want to improve. And exercise can do it. It'll help them improve across the board."

For building cognition, Sudoku is a shovel and exercise is a bulldozer.

When Kirk Erickson, then Kramer's postdoc student and now a psychologist at the University of Pittsburgh, put a group of seniors on an aerobic exercise program and measured their hippocampal growth at six months and twelve months, the surprise was that the neurogenesis didn't taper off. The hippocampus kept expanding like a balloon. *There was as much growth in the second six months as the first.*

What's most surprising is the sheer variety of ways that regular exercise leaves its signature on the brain.

It appears, for example, to make us better multitaskers. Recently, a group of 7- to 10-year-olds were invited into the Cave at the Beckman Institute and run through the same street-crossing task that Olga performed. It turns out that modern kids with higher fitness levels—who, by the way, show no aversion to being hit by virtual cars—more effortlessly multitask than their less-fit peers, and so avoid collisions. It didn't matter whether the higher-fit kids were talking on the phone, listening to an iPod, or walking undistracted: they crossed the street more successfully. Better still: "The fitter children didn't walk any faster," says Laura Chaddock. "The data suggest that they may *think* faster."

There's even evidence that regular exercise makes us more creative. In one 2007 experiment, subjects who had run on a treadmill for thirty-five minutes at 60 to 70 percent of their max came up with more out-of-the-box solutions to problems. "Sometimes I runs and thinks, and sometimes I just runs," the baseball legend Satchel Paige almost said. Running and thinking is better. With exercise, the whole brain blooms.

Exercise is a wondrous paradox, in the sense that it's both a stressor (as we saw in chapter 2) and a *stress reliever*—maybe the best one known. It grows the hippocampus—the brain region that senses stress and reacts to it, marshaling a stress response.

So quite apart from producing happy neurochemistry that bleeds off stress in the short run, exercise renovates the part of the brain that actually manages stress.

When Olga stumps for the virtues of exercise, she downplays that first part of the old adage, "It adds years to your life," and instead leans on the second part: "It adds life to your years." Exercise makes us happy, in a deep and lasting way. It chases away the blues and keeps them away. (A 2000 Duke University study found exercise to be better than Zoloft at treating depression.)

There appears, indeed, to be something uniquely laid-back about the exercise-built brain. Recently Princeton researchers built new neurons in the hippocampi of sedentary mice by exercising those mice, and then they looked closely at the new neurons they had created. There was something different about them. They showed a dampened response to stressful situations. The brain of a person who has diligently exercised for many years—a brain thus full of neurons built from sweat—seems to dial down the level of chatter in the stress circuits.

It's hard to nail cause and effect here, but you can see the probable results of this long-term calming effect at any masters track meet. There is an almost Buddhist serenity about many older masters athletes that you notice the moment their race ends, and the default personality resumes. (It's the disconnect between the beetle-browed competitiveness on the track and the easygoingness off it that I find most striking about a lot of masters athletes, Olga among them.)

It's not yet clear what *kind* of exercise is the best brain food. Thus far, the case for aerobic exercise is strongest, if only because far more research is behind it.

Aerobic exercise has been shown to boost our central command functions, such as the ability to think critically and deal with ambiguity. Around age 40, our neurons start losing their insulation and other changes befall our brains. But aerobic exercise re-insulates the axons of those brain cells, boosting processing speed and making the connections more reliable. We know, too, of aerobic exercise's fantastic cognitive multiplier effect. It "sparks production of neurotransmitters and neuro-trophins, and creates more receptors for them in key areas of the brain," as Harvard's John Ratey puts it. And it turns on genes that keep that positive cycle spinning.

We know less about the brain-building potential of resis-tance training. (Why? Because animal research always precedes human research, and it's a lot easier to put a mouse on a run-ning wheel than to get him working out with mouse-sized ket-tlebells.) But there's evidence that weight training does some of the same kinds of things for the brain as aerobic exercise does—things like stimulating proteins called neurotrophins, which signal brain cells to survive and reproduce. And it may do some brain-building things even better than aerobic exer-cise does. Teresa Liu-Ambrose, director of the Aging, Mobility, and Cognitive Neuroscience Lab at the University of British Columbia, found that when older people lifted weights they improved their executive control—things such as scheduling and planning and dealing with ambiguity—even better than the group who did aerobics alone.

Some researchers believe that if you can get the heart pump-ing while performing more skillful, complex movements, you truly have a recipe for a superpower brain boost. The research group at Beckman is currently testing dancers to see if their brains are yet more agile than runners and walkers who are other-wise similarly fit.

In this respect, Olga may have stumbled upon a perfect for-mula. Her eleven track events move her body through space in almost as many ways as a body *can* move through space. At the same time, there are explosive anaerobic demands. And at the end of it all someone puts a gold medal around her neck and tells her she's wonderful.

There's no better brain meal than that.

EXERCISE isn't the greatest correlate of longevity. By far, the greatest predictor of longevity is *means*—the good fortune to be born into a comfortable, well-educated family in a developed country. But exercise, increasing evidence suggests, is number two. It may matter more than diet or occupation or even genes. Ralph Paffenbarger, a Harvard epidemiologist and one of the first to establish the link between exercise and longevity, crunched the data from more than fifty thousand subjects to arrive at a rule of thumb: every hour of exercise we do amounts to two hours tacked on to our life span. (The formula is a blunt instrument. It now appears, as we'll see in a moment, that all exercise is not created equal.)

Exercise buys us a chance at long life by lowering the risk of a variety of ailments—heart disease chief among them. It can reverse the effects of a genetic bad hand—by, for instance, switching off genes that predispose you to obesity. It seems to slow aging through such measures as promoting the growth of stem cells in muscle, expressing genes linked to longevity, and even lengthening telomeres. That doesn't mean the lifelong exerciser has a hope of reaching 125. But it does mean that he or she is just younger, every step of the way, than someone who doesn't exercise.

So commonsensical is the advice to keep moving as we get older that we forget how new it is. Until 1972, when an influential paper upended the paradigm, the "disengagement" theory ruled the day. What old folk ought to do as they nosed into the golden years was . . . nothing. They should quietly, inertly, withdraw into themselves until death came for them in a taxi. (The Germans have a word for this: *wohlverdient*. It means the "well-earned" rest of rocking-chair-bound retirement.) Today, any doctor who counseled that would probably be reported to his or her state medical board. The new line is that "our evolutionary past designed us to be active and fit until we drop dead," as the UC Irvine evolutionary biologist Michael Rose puts it.

Moreover, there's increasing evidence that when you add exercise to *anything*—from meditation to a healthier diet—you get "synergy effects." Exercise just makes every good habit you have more potent.

Here is the enigma, though. Olga, as a regular exerciser, is one among a couple hundred million in the tent. But there is only one Olga. So if exercise is indeed one of the secrets of her youthful vigor, we need to zoom in a bit tighter on this picture, to find out why in her case it has had such a profound effect.

Get Out of Jail Free

In his guest lecture to the McGill kinesiology students—the one he'd opened with the weight lifter in the high shorts—Russ Hepple put up another slide.

Here was a 70-year-old cyclist. Another fine specimen. But this man had a different story than the previous guy who had worked out hard at a high level all his life. This man had been a runner in college, and then, like so many of us, let his body slide into desuetude. For close to four decades he did nothing.

"When he hit 62 he thought he'd get into cycling," Hepple said. "The next year, at age 63, he started competing."

Now he is 70. "His max heart rate? 166 bpm. VO_2 max is 4.7 liters per minute, or 60 milliliters per kilogram per minute."

Hepple looked around the room. Did the students understand what those numbers meant? His VO_2 maximum—a measure of oxygen delivery and a good shorthand for aerobic fitness level—is *60*. "Most girls in this class will be around thirty-five. Guys? Maybe forty.

"So this is astonishing," Hepple said. "He was a couch potato for forty years and then started back." It was a parable for the body's remarkable ability to find the old groove, to erase the debts of decades of bad lifestyle choices.

If there's an emerging big story from the front lines of exercise science, this is it. Exercise is the boomer's Get Out of Jail Free card. For years it was assumed that the crucial exercise window was in childhood, or perhaps teenagehood—and that there was little hope of any dramatic benefits beyond age 50.

We now know that's not true. People who start exercising in midlife can make massive gains. They may not hit quite the same high-water mark as those who were fit from the get-go, but they actually travel a greater distance. They improve more on a host of measures, including aerobic capacity. Some studies show a 70-year-old can improve his or her VO_2 max score, through training, as much as a 21-year-old can.

Why? One theory is that these other mechanisms somehow compensate for the lower cardiac output that's inevitable as we age. The body of the older jock says, "Okay, where can we gin up performance gains here?" Exercise seems specifically to bolster the parts of us that normal aging erodes. It may be nature's way of sandbagging the parts of the dike that are most likely to be breached.

Think back to the brain region we spoke of earlier, the anterior hippocampus, best known for its role in memory. It's famously vulnerable to the ravages of time—shrinking at a rate of about 1 percent a year after age 20. But when scientists put both young mice and old mice on treadmill programs, and then look at their newly exercise-rebuilt brains, the older rats' brains steal the show. "The absolute number of new neurons is greater for the younger animals," notes Art Kramer. But the *proportion* of new neurons is greater for the older animals. "So you're really getting a bigger bang for your buck when you're older."

The minor miracle here is, you can introduce exercise at any point, right up into very old age, and "completely reverse any decline you've had," notes Justin Rhodes.

The Distaff Factor

Be doubly celebratory at this point if you're a boomer or older and you're a woman. Because in some ways exercise offers you bigger payoffs.

The psychologist Kirk Erickson, now at the University of Pittsburgh, found that women get a slightly bigger cognitive lift from aerobic exercise (that is, they grew more brain cells, faster). Laval University in Quebec suggests that the dose effect of that cognitive benefit from exercise—the more you work out the sharper you get—is greater for women. Yet another study, by the psychology department of Appalachian State University in North Carolina, found women reported a significantly bigger energy lift from both cardio and resistance exercise than men did. The boost in energy may be especially welcomed by postmenopausal women—a kind of compensation for the hormonal swoon.

The little old lady, it's starting to look like, has been miscast in our culture. She has been consigned to the role of pie maker and child minder when she should be helping raise the barn. She responds to exercise like nobody's business. There's evidence that a woman can more safely push herself at the gym deeper into old age than a man can. For while it's true that women are endowed with less muscle than men, it is in some ways *better* muscle—that is, muscle more resistant to breakdown. (One reason may be that estrogen seems to promote muscle repair and maintenance, Wilfrid Laurier University kinesiologist Deborah Enns found.) Get a group of men and women in the gym for high-intensity weight training and then get a microscopic look at their muscles afterward, and you would find something remarkable. The exercise has grown age-defying muscle stem cells in everyone. But, University of Maryland kinesiologist Stephen Roth found when he tested men and women, young and old, the response was most pronounced in the *older women.*

All of this may explain why, on the last day of a big masters track meet, the lines for the medic tent look like the opposite of the lines for the public restrooms: more men than women. You could say the guys have been writing checks that their bodies can't cash.

Ninety percent of centenarians are women. No one knows for sure why. But the fact that women generally hold off life-threatening illness about a decade longer than men do, and they are more likely to survive it, tips conjecture toward a conclusion that exercise science is now supporting: there is a resiliency in little old ladies that puts them in rare animal company—up there with tortoises and terns.

Advantage, Olga: times two.

Sweat Equity

There is a third consideration. Turns out the rigor we apply to the exercise we do counts, in ways scientists are just beginning to understand. In a nutshell: intensity matters.

Any running addict can tell you that really going for it is *worth it,* psychologically. The discomfort is a chit you can cash at the pay window thirty-five minutes from now, as endocannabinoids flood the system and worry evaporates. Every minute you push yourself beyond your comfort zone is redeemable as a unit of bliss, and you can coast at least a whole day on it, as locked in as a Zen master on Ritalin, before it starts slowly wearing off.

The good news is that the psychotropic effect of exercise does not diminish as we age, according to the Swedish School of Sport and Health Sciences. The same level of endorphins is produced, the same high is felt. Thus does exercise loom, for seniors, as a tantalizing outlet—a reliable pleasure source in a world of diminished pleasures, the keys to the hooch cabinet. The trick is, you have to get to a certain fitness threshold before you get to touch the runner's high. You have to work your way up. ("And it really is not much fun at the start," concedes Arthur Kimber, the 80-year-old British masters miler. "You've first got to get fit to be able to run. And then you've got to run to be able to train. And then you've got to train to be able to race." The prize of the competitor's high is hard won.)

But something more is happening in the brain than the trippy payoff we feel when we really bear down.

"Why is hippocampus growth so quantitatively related to *intensity* of activity? That we don't know," says Justin Rhodes. "But the relationship is clear. When you exercise the cells in your hippocampus are firing together in synchrony, in proportion to how active you are. If you could put electrodes on Olga's hippocampus and observe her on a treadmill as she ran faster and faster, you'd see it. We're not just talking about a few cells; these are large numbers of neurons firing together that produce electrical potential you can read outside the scalp."

Intensity concentrates the physiological benefits of exercise. Research on subjects who were guided through short, periodic blasts of cycle exercise in Martin Gibala's lab at McMaster University in Hamilton, Ontario, suggests we can get by on *seven minutes of exercise a week*, if that exercise is intense enough.

There seems to be a threshold beyond which exercise does

something completely different to us—something very much like rolling back the odometer.

Recently, Mark Tarnopolsky, a pediatric medicine professor from McMaster, got his hands on a few breeding pairs of a very special strain of mice. Scientists had knocked out an important gene. In effect they had vandalized these mice's genetic photocopier, so every cell division would make increasingly flawed copies of their mitochondrial DNA.

The change was quickly obvious. These were Benjamin Button mice. By a few months old they were becoming gray and wizened and weak, with shriveled sex organs and sclerotic hearts and failing hearing. They looked to have a foot in the grave.

"The mice were developed to mimic mitochondrial disease," says Tarnopolsky. "But when we saw the animals we thought, holy crap, they look like they're old! They have all of the features of normal human aging."

Meanwhile, Tarnopolsky had put a separate group of these genetically modified mice on an exercise program while they were young. Three times a week they ran on the wheel for forty-six minutes at a moderate clip. (About a nine-minute-mile pace for a person—which approximates the routine for a lot of us.)

Eight and a half months later the results were staggering. Tarnopolsky published them in a 2011 paper in the *Proceedings of the National Academy of Sciences*. The running mice were rejuvenated. A sleek, dark coat had replaced the sad patchy fur. Emaciated muscles had plumped back up, shrunken gonads bloomed again, and unhealthily enlarged hearts returned to normal form. At the cellular level, mitochondrial DNA mutations had stopped, and new mitochondria appeared. Antioxidant enzymes appeared to mop up the free radicals that the exercise itself had produced. In just about every measurable way, exercise reversed the signs of premature aging in these mice. They were now, Tarnopolsky says, "indistinguishable from wild mice who did not have the gene mutation."

Crucially, it was the effort level of the mice that promoted the astounding recovery, Tarnopolsky believes. They were going for it, relatively speaking. It wouldn't have worked if they weren't. A certain critical intensity threshold seemed to be required for these startling cellular changes to happen.

To test what that threshold is, at least in mice, Tarnopolsky conducted a "voluntary running" experiment. He gave a group of mice access to the wheel every night. Mice like to run, but some like it more than others. Some mice ran a lot, and after about two kilometers those mice began showing "a very significant rescue of their mitochondrial capacity," Tarnopolsky says. (Beyond that point the benefits tapered off.) But some slacker mice ran only about five hundred meters, "and they derived almost no benefit."

Studies in humans show a similar difference in gains according to effort. In one study—by exercise physiologists at the Manitoba Institute of Child Health, published in the *Archives of Pediatrics and Adolescent Medicine*—a group of adolescents who swam or played soccer for just ten minutes a day got fitter and healthier than another group doing almost twelve times that amount of "light activity"—moving but not exercising. Now, hard exercise for kids is one thing. The degree to which it's appropriate for older people is an evolving issue. Indeed, just three weeks before Olga was first scheduled to step onto a treadmill at McGill for aerobic testing, Tanja Taivassalo received an e-mail that her project was on hold. The McGill ethics board had called a halt. It was concerned about redlining a nonagenarian in its facility. It seemed risky. And anyway, Olga couldn't consent because the forms only went to age 80.

To Taivassalo, this hitch was somehow a fitting metaphor for the problem she was tackling: many people fail to realize that the benchmarks defining "old" have moved. Providing they're healthy and fit, even very old people are capable of pushing themselves aerobically quite safely, increasing evidence suggests. And there's no substitute for what stress-testing aging bodies can tell us about human physiology. Taivassalo prepared a brief, presented her case before the ethics board, and sewed it up in fifteen minutes.

"Part of the challenge is the mind-set or dogma that we need to slow down as we get older," says Scott Trappe, director of the Human Performance Lab at Ball State University in Indiana. For example, the belief that aging joints and tendons can't take real weight training is dead wrong; real weight training is what might just save them. Seniors can work out less frequently, Trappe reckons, as long as they really bring it when they do. It's all about intensity. How much intensity? "At least 80 percent of your max," he says.

The track events Olga undertakes stress her body in exactly the right way, which is to say, they make heavy demands for a very short time. Her body mounts a ferocious stress response and then rebounds to rest. Hormones called amines flood the system and then quickly clear out. And with the rest comes the adaptation. (This is why some experts believe ultramarathons or even marathons do participants no favors. They put too much stress on us for too long.)

Let's be clear that "light activity" is still vastly preferable to no activity; and for people who aren't particularly sporty it's a perfectly decent prescription for maintaining basic health. But "there definitely is some intensity threshold where exercise starts becoming *really* beneficial," says Tarnopolsky. A growing consensus of opinion says that to derive maximum benefit from exercise you need to sweat. You need to challenge yourself enough to provoke an adaptation response. Harder is better . . . up to a point.

It now appears that there is a limit. This is the takeaway from two big recently published studies, and it matters to a whole generation that took up jogging in the 1960s or '70s, became addicted, and never eased off the throttle. Just as you can underexercise, you can overexercise. Do too much, too hard, for too long, and you start doing more harm than good. Heart scarring or a heart-rhythm disorder can result.

"There's a fivefold greater risk of atrial fibrillation in people who have done long-duration or high-intensity exercise for a long time," notes Tarnopolsky. The culprit, some suspect, is oxidative stress. The symptoms may not show up for a long time—perhaps a couple of decades. "If you're starting in with extreme sports when you're young," Tarnopolsky says, "it's going to take a while for that degenerative process to manifest itself." Actually, he himself fits the profile. "I went almost twenty years without taking a day off," he says. His own AFib was diagnosed at age 46.

Olga mostly gets a pass from overexercising concerns, for a simple reason: she is not an extreme athlete. I will say that again: she is not an extreme athlete. It may seem like she is, given her explosive intensity and the ridiculous breadth of her event menu. But she has never really pushed herself aerobically for long stretches. She probably never redlined her heart for more than an hour in her life. Her adrenal glands have some mileage on them still.

Is she addicted to exercise? I would say she's addicted to the trappings of it: to the success, to the winning, to the meaning it gives to her life. But she's probably not physiologically addicted in the way that causes so many lifelong runners to blow past the caution flags their own bodies are raising.

Exercise is a wide-open door. It offers its benefits to just about everyone willing to put in the work—and indeed it seems in a sense to offer compensatory gains to the aging and the weak and the late-to-the-party. The little old lady who has the drive and opportunity to exercise hard, and the luck not to get injured doing it, is presented with an astonishing portal to good and lasting health. This stands as evidence against the idea that Olga is a freak of nature. In other words, there could be a lot more Olga-like specimens, if circumstances lined up for them as they did for her.

All of which enlarges the mystery.

Here's where the "transhumanist" tribe, banking on technology to trump their mortal limitations, starts looking to the future for clues.

But Olga and I decided to look the other way: deep into the past.

5

An Evolutionary Theory of Olga

The Test of "Species Strength"

IN A HANGARLIKE building in an industrial strip in East Vancouver, a gentleman called Sheppy stops writing on the blackboard and fixes Olga and me with a purposeful look.

"My job," he says, "is to make you harder to kill."

Sheppy's a tall and sinuously muscled fellow, with rimless glasses that lend him a Clark Kent–ish air. He's one of the top trainers here at the flagship Canadian location of CrossFit, a chain of bare-bones yet high-concept gyms. Really, CrossFit is a kind of anti-gym. If you wandered in by accident, it would not be immediately apparent what's going on here. A black-box theater company? A twelve-step group? There's no fancy fitness club equipment: no treadmills, no StairMasters, no step-in weight machines. Just wooden rings hanging from straps and, on the gym-mat-covered floor, some barbells and hand weights. A fine dusting of grip chalk. A sign on the wall says: "No shoes, no shirt, no problem."

I'd asked Olga here today because the Way of CrossFit feels so complementary to the Way of Olga that the two needed to meet.

CrossFit was developed in the 1990s by a former gymnast named Greg Glassman, who concluded that a lot of people are exercising wrong. In his view we've lost sight, in the age of pretty-boy abs and the fetishization of MVO_2 scores and such, of what it really means to be fit. "Fit" in not one or two but ten domains: stamina, strength, flexibility, power, speed, coordination, agility, balance, accuracy, and respiratory endurance. Fit like a fireman or a cop or a burglar. Fit in the sense that you can snap into action out there in the world, whatever danger comes at you, whatever opportunity presents itself. It's the kind of fitness that, two million years ago, would have kept us alive to hunt and fight and *schtup* another day. By "harder to kill," Sheppy explains, he means by any conceivable threat, "be it a disease or a grizzly bear."

It's all crazy-macho stuff. But Olga seemed immediately to get it. Because what she does is pretty macho, too. She runs, jumps, heaves a rock, throws a spear. She does for fun and fulfillment what we once did for survival. Those challenges, plus a small group of others, have become enshrined as a kind of test of species strength.

"The reason some sports are in the Olympics is that we've been doing them forever," Sheppy says. "Swimming? We came from amphibious creatures—it's part of our genetic memory. Rowing? There's evidence that early man was paddling canoes on the West Coast here while the ice sheets were around. These are *things your body wants to do.*"

Could this be part of the secret of Olga? That she is doing, and has done for her whole life, exactly what our body is built for?

That's really the nut of "evolutionary fitness"—and the whole paleo movement more broadly.

The theory goes something like this: many health problems of the present day are due to a mismatch between our genes and our environment. Most of our major genetic adaptations occurred one million to two million years ago, in the so-called Environment of Evolutionary Adaptiveness for humans, and any changes since then have amounted to proofreading corrections in the galleys. Our genome has barely changed a whit in ten thousand years. We are Stone Age people living in a world our bodies can't quite grok. From the food we eat to the (lack of) sunlight we get to the kind of exercise we get (or don't), modern life is alien to our design specs.

Certain diseases of the body—such as obesity, heart disease, diabetes, hypertension, and some cancers—and certain diseases of the mind, from depression to ADHD, were all but unknown in our ancestral past but have gone epidemic in just this past century. No one knows why. But some eloquent defenders of the mismatch theory, such as University of Missouri physiologist Frank Booth, are confident that the disconnect between design and application explains it. "We're maladapted for our genetic instructions," he writes, "leading to abnormal gene expression, which in turn frequently manifests itself as clinically overt disease." We're using this machine called the human body in ways that would void the manufacturer's warranty.

Most of us are, that is. Olga isn't.

If you buy the mismatch theory, then the key to robust health over the long haul is to try to live in a way that shrinks the mismatch as much as possible.

Cue Olga. She is more behaviorally paleo than anybody I know. Diet-wise she's a little off the formula, as strictly defined. (Though she likes her lean meat and she likes it rare, she eats more carbs than those diets allow, and she's more of a grazer than many of our early ancestors were.) But in the way she uses her body she's textbook.

If you were a Paleolithic Everyman, this was your life: chores—of the hewing wood and hauling water and butchering meat variety—lay in front of you from dusk to dawn. A Paleo-lithic woman's "exercise routine" would look like hour upon hour of "arduous digging, climbing, bending, stretching and carrying heavy loads back to camp," notes James O'Keefe, a cardiology professor at the University of Missouri at Kansas City who has written extensively on the topic. And she often had the added burden of carrying a kid.

Almost nobody in any Western culture gets this level of daily exertion nowadays. The Amish, unseduced by the labor-saving benefits of modern technology, perhaps come closest.

But over her lifetime Olga has been on the move more than most, starting as a toddler on the Saskatchewan farm.

She was milking cows as soon as she was old enough to carry a bucket at dawn. She was the washer of fifteen sets of clothes. She'd scrub them on the washboard and hang them on the line where in winter they'd swing, frozen like boards in the wind. Then she'd commence ironing.

She baked bread—twelve loaves at a time, twice a week—and hatcheted chickens. She routinely traveled the two miles from the homestead to the nearby town of Smuts, along the railroad tracks, carrying pounds of butter and dozens of eggs to sell at the grocery stores.

In the grain fields, the binding machine would leave a row of sheaves as thick as barrels, and she would pick up those sheaves and carry them, one under each arm, then stack them to dry so they'd be ready for the thresher.

For two weeks each year she joined the rest of the family picking rocks. Anything smaller than a cantaloupe you could leave; anything bigger you schlepped away before it could be caught in the plow. For two weeks you cleared the land. And lo and behold, in the spring, the field was full of rocks again, heaved up through the field by the winter's frost. So you did it all over.

She walked two and a half miles to school after those chores were done, making sure to arrive early enough to get a pickup softball game going with a rag-stuffed ball. Sometimes she and her siblings took the team on the road, walking almost seven miles to another school to play their kids. In they would saunter, a group about the size of a Paleolithic hunting band, for the purpose of crushing an opponent.

Home by moonlight. Rest, then repeat.

Life hardly got easier for Olga after she left the farm: teaching in a one-room schoolhouse, then fleeing that horrendous marriage and raising two girls on her own while attending night school. Then adult softball, and then track.

For paleo men and women, the drudgery of the daily busywork was broken up by some really intense stuff, and your system had to be ready for it. So, too, with Olga. On the track, her favored form of training has been intervals—"slow for a minute, fast for a minute, slow for a minute"—the better to prepare her for the 100-meter and especially 200-meter dash. There's a reason no one else in the world her age still high-jumps and long jumps. That's off-the-charts intensity for a 90-year-old body.

The Brass Ring

Back in the CrossFit gym, we put on our running shoes in the transition area: from classroom to workout zone. The midday crowd was thin.

Honestly, I had expected a somewhat different clientele. You hear of the paleo-cultists from central casting: guys who grow beards to trap the musk of all those growth hormones they're secreting (to lure the ladies and intimidate the men). People who fast before working out, to stir ancient memories of the ravenousness that preceded a mastodon takedown.

In fact, there's no one like that here. Just two cops, a dock-worker, a bartender, and a stay-at-home mom. Olga spots some-one she recognizes: a petite, friendly woman named Jennifer Schutz. Schutz is a running coach for the University of British Columbia track team. She's here to do the "daily"—a workout chalked on the wall like the specials menu in a restaurant with revolving chefs. (You could go to CrossFit for three months and never repeat the same workout, which is precisely the point: to surprise the body and force adaptation.) Strength training has made her both stronger and faster—shaving min-utes off her 5K time, Schutz said. She goes over to the high bar and starts doing pull-ups, and she is still going, with no end in sight, when Sheppy summons us over to get a read on our fitness level.

We sit on the floor and start with some dynamic stretching. Olga is not *super* supple, but she is, I would say, limber. She manages the routine no problem.

I can't even get into the start position. If this is a clinic on "functional movement," I am the "before" picture. I am a guy who brings a full-sized chair on hikes because I can't even sit on the ground. That comes from not stretching for thirty years. (My theory of not stretching is that if you keep every part of you tight, like the rigging on a ship, then nothing can move in a way it's not supposed to and you don't get injured. This proves effective. You can barely move, but you rarely get injured.)

Sheppy hands us each a big leather medicine ball. He will time us while we run out of the gym, down the alley, up a hill, around the block, and back, carrying the ball via whatever method we hatch on the fly. Mine weighs about twenty pounds. You can imagine an evolutionary context: I have to get my kill back to camp before the family starves. Olga's ball is lighter—

around the weight of a newborn baby. She is hustling her grandchild away from a cave bear, maybe.

This for me was part of the intrigue about today's visit. I was curious to see what kind of program they would devise for Olga—that is, how different from the men, how different from the young. I figured an evolutionary fitness program would recognize the different roles a middle-aged man and an old woman would have played in the tribe, back when.

But it doesn't, because there's evidence that those roles weren't so different after all.

Paleolithic granny was a far cry from the frail, stay-at-home, fire-tending figure we imagine. There's evidence she was a whirlwind of productive vitality, roaming wide and quickly to forage nuts and tubers for her grandchildren. She was every bit as physiologically adapted to strenuous labor as the men. Natural selection favored Olga-like stamina and longevity in grandmothers—because the longer cave granny lived and the more energy she had to help out, the better the chances of the survival of her genes through her grandchildren.

Indeed, in the mid-1990s, when then–Harvard anthropologist Frank Marlowe started visiting the Hadza people of Tanzania's Olduvai Gorge—one of the last remaining true hunter-gatherer societies and perhaps the closest thing to a genuinely Paleolithic lifestyle today—he found communities teeming with Olga-like female elders. Just about every nursing mom had a hardworking granny helping her out. The old women's experience makes them cannier foragers than the younger Hadza, and they are actually more industrious—indeed, elderly Hadza women brought more calories back to camp than any other group. Grandmothers are the engines driving the survival of the family.

As it is now, in Olduvai Gorge life, so it was a million years ago. Little old ladies were endowed to handle as wide a repertoire as the men, and that versatility is in them still.

"We've used our same routines for elderly individuals with heart disease and cage fighters one month out from televised bouts," Glassman once told a reporter. "Of course, we can't load your grandmother with the same squatting weight that we'd assign an Olympic skier. But they both need to squat."

There's a lot of squatting in CrossFit.

The idea is to get strong from the inside out, to forge a strong core and "posterior chain," on which all skilled movements depend. The idea is to do them not slowly with a whole lot of weight piled on you, but steadily, repeatedly, mixing in other exercises and returning to the squat, until your heart is hammering in your chest.

Nobody has ever surveyed the percentage of people who puke after their first CrossFit class, but it's high. "If I told you to drop right now and give me fifty push-ups, you'd grunt, you'd spit, you'd swear, you'd feel like you want to die," Sheppy told us. "Well, *I want you adapted to that feeling.*" Most of us never even approach our physical potential because we don't push ourselves enough to get the adaptation cycle going, Sheppy says. "It's at the fringes of our ability that we make the most gains." There's an American-style appeal to CrossFit training. This, writ large, is the Cult of Improvement. We try, we fail, we try harder and fail better, we make slow but steady progress, and in that way we move forward, farther than we used to go, and then farther again.

Next it's pull-ups—as many as possible at one go. (A stretchy rubber band is fitted around Olga here to take some of her body weight—which doesn't seem fair at all.) Then it's push-ups— again, as many as we can do till we can do no more. Dips: same deal.

For some reason, after all this upper-body work, my *thighs* are sore.

"Well," says Sheppy, "How often do you push yourself to your absolute maximum? You've shocked the whole system. Your body is going, 'What the fuck?' "

There's a certain strange appeal in inhabiting a world, however briefly, where people talk like this. CrossFit promises to get the juices flowing again in midlife meat-puppets who have lost touch with their animal selves. If you're feeling pain, you are at least . . . *feeling*. The high comes in finding a new register of sensation, good and bad. Enervated, alert, hungry, achy, tranquil, horny. *Young*.

"CrossFit is the brass ring," Sheppy says. "We've figured out how to boost three essential growth hormones. I tell people I'm a drug dealer. No one can touch my stuff. Testosterone, HGH [human growth hormone], ILGF [insulin-like growth factor]. In the beginning I can only give you little doses. But each time you come you get a little more. This is it—the bones of human performance. Put a stopwatch on it, you've got a sport. Add a beer, you've got a community."

Olga listened attentively to Sheppy's soliloquy, but failed to be as stirred by it as she might have been because she'd forgotten her hearing aid at home and so missed a bit. But at one point Sheppy caught her attention. He had just boiled down the CrossFit philosophy to "more work in less time," which is one definition of intensity. "What we're doing here is teaching your body to explode," Sheppy said.

"My body doesn't explode!" Olga replied.

The thing is, though, it does. Relative to other 93-year-olds, Olga is a dangerous ordnance. Save the 200-meter dash, the event she finds most torturous by far, all of Olga's events make anaerobic, not aerobic, demands. Close to 100 percent effort for just a few seconds. The one thing Olga lacks is aerobic endurance—that longer but lighter-loaded effort. The degree to which that matters approaches the heart of the controversy about what our bodies were—and are—built to do.

Born to Run vs. Born to CrossFit

No question about it, our genes designed us to be action figures. We function optimally when we get tons of exercise—way more than most of us do in our modern desk- and car-bound lives. The question is whether Olga's robustness is perhaps linked to the *kind* of exercise she does.

Here's where the debate heats up.

Olga is obviously well adapted to what, as Sheppy puts it, "our bodies want to do." But is she better adapted than someone like, say, Ed Whitlock, who can make a contending claim to be the best masters athlete in the world?

Whitlock, 81, is a marathoner from Milton, Ontario, known not just for his records but for his training regime: daily three-hour, turtle's-pace runs around the local cemetery. Whitlock calls to mind a somewhat different evolutionary paradigm, one made famous by Daniel Lieberman in his 2004 paper with University of Utah biologist Dennis Bramble (and popularized in Christopher McDougall's bestseller *Born to Run*). Lieberman argues we are built to motor along like Energizer bunnies. For hundreds of thousands of years—at least until we figured out how to kill at a distance with spears and arrows—this was our advantage: we're not faster than most other mammals, but we have great endurance and specialized anatomical traits to make our feet springy and our heads stable and our bodies cool over the long haul, so we could outlast our prey and club it dead with a rock when it collapsed from heat exhaustion. This helped us pull away from the Neanderthals. "Persistence hunting" was our ticket to the top of the food chain. Among their evidence, Lieberman and Bramble point to twenty-six features early *Homo sapiens* evolved specifically connected to running, including short toes and big butts. Even the vaunted "runner's high" is thought by some to be an evolutionary fillip—nature's reward, quite apart from the dinner now before you, for the sustained and fairly intense effort it took to see the hunt to completion.

No one seriously disputes that we were born to run. (And jump and lift things.) Now we're just haggling over distance, over the point where aerobic exercise shades from helpful to harmful—where it stops building you up and starts breaking you down.

It's not at all clear that revving our hearts for much more than an hour at a time is either good for us *or* part of our evolu-tionary story. A credible counterthesis holds that competitive marathoning isn't really very much like persistence hunting—which involved slow jogging, walking, and stopping to rest.

Are we natural-born marathoners or something closer to natural-born gymnasts—or, because we hunted in packs, team sportsmen? (Because it's unprovable, the debate invites all kinds of fun and geekish fantasy-league-style speculation. "Put a group of NBA players and a group of marathoners out there on the veldt," one blog poster posited, "and let's see which brings down the most antelope.") Which side of the debate wins matters because it will help shape the future of fitness: whether we aim for something like Ed Whitlock's approach or something like Olga's. Long and slow and relentless; or short and intense, over and over.

In some ways it's an unproductive question. Because the emerging consensus is we evolved for both endurance *and* bursts of power. Which one you needed most, notes William Meller, a physician and professor of evolutionary medicine at UC Santa Barbara (and author of the book *Evolution Rx*), depended on where you lived. Out of Africa, where there were few places for a hunter to hide, came the distance runners, with stamina their trump card. "But in many more places our ancestors used stalking and trapping. They used their brains. They developed the skills to sneak up on something and whack it on the head." In other words, we needed both—the long and graceful and the quick and dirty—and we still do. In terms of the ways they create energy, the suite of hormones they release, and the specific health benefits they promote, they're complementary modes.

Even Dr. Ken Cooper, who coined the term "aerobics" and championed long runs and rides, now believes man cannot live on cardio alone. We need resistance training, too. Even as we get old. *Especially* as we get old. Endurance aerobics aren't much help when muscle mass shrinks (making us weak) and fast-twitch fibers vanish (making us slow) and usable testosterone wanes (making us, well, *unvital*). Only intense-ish resistance work seems to do much good on these scores.

"Olga's not doing the same thing day after day, and that's key," says Mark Tarnopolsky. "Our research shows that if you mix in endurance activity with strength activity, the two in combination are what probably allow people to continue to train and to get the maximal benefits as we get older. If you just do one you compromise the other. So if you just run all the time—and scientists found this when they tested top orienteers and cross-country skiers who were former world champions— their strength was no better than sedentary individuals. So it's critically important to mix it up."

If you couldn't be talked out of choosing one over the other, you'd want to be clear about your goals. If palliatives are what you're after—being blissed out, sleeping better, enjoying a beanpole-slim profile, and maybe bucking up your short-term memory—endurance aerobics could be just fine. But to actually *age more slowly,* you might well want to tip toward shorter, harder, Olga-style workouts.

That has always been Earl Fee's line. Fee is a legendary Canadian sprinter and middle-distance runner, current world 800-meter champion for men over 80, and holder of more than a dozen world records. "The thing that keeps you young is the intensity," he says. "I call it SFA—short, fast, aerobic interval training." Fee worked as an electrical engineer, designing nuclear reactors, and he views the human body as a machine that runs better, and longer, if you methodically fuel it and maintain it correctly. Fee scours the science journals and distills their findings in books such as *100 Years Young the Natural Way,* which he sells out of a suitcase at track meets. Fee himself takes thirteen nutritional supplements, and tries to eat a ratio of 80 percent alkaline foods and 20 percent acidic foods. I once saw him divide his dinner onto two plates—one each for meat and potatoes. He has strong ideas about never mixing flesh foods with starch, as this is a hard-to-digest combination. But mostly, Fee, an intense man himself, is fully committed to the idea of intensity: High-intensity training to confuse the muscles and open the lungs and promote adaptation. High intensity to tip the balance of muscle fibers toward fast-twitch, and crank up the production of human growth hormone—that elixir of youthfulness you'd pay a fortune for on the street. Fee routinely pushes his heart rate beyond 180 beats a minute— which is a little like inflating the tires of your thirty-year-old Chevy to fifty pounds per square inch: terrified onlookers watch through their fingers.

It's fascinating, in this vein, to compare Earl Fee to Ed Whitlock. They are North America's two best octogenarian track athletes. Fee is far and away the fastest 80-year-old half-miler in the Western world. Whitlock is far and away the fastest

80-year-old distance runner in the world. (Two years ago he ran the Toronto marathon in 3:15 and change. That age-grades to 2:02—the fastest marathon ever run.)

But if you were looking for secrets of success the two men have in common you'd come away scratching your head.

Whitlock, who, coincidentally, was also an engineer, trains at a metronomic slow pace, for hours. He seems to subsist mostly on grilled cheese sandwiches and coffee. He does no hill training, no cross-country. In theory and practice, Whitlock basically contravenes everything that Fee so fastidiously promotes. They are both regarded as something like gods in their circles. (Olga counts them both among her sports "heroes," a list contained on one hand.)

But there's one obvious difference between them. Fee looks freakishly young. He prompts double takes when he removes his shirt—it's a wrestler's body. He is a living advertisement for his self-professed strategy: to "age more slowly than my competitors."

Whitlock, on the other hand, is in no danger of being pestered by officials for his birth certificate. Delicate veins in his eyelids spider through paper-thin skin, and hound-dog bags rest beneath, giving him a kind of default sadness. As he grinds out those slow laps around the local cemetery, watch dangling from his broom-handle wrist, the impression, to passing motorists, is of a terminally restless white-haired ghost.

Whitlock's sheer mileage may be wearing down the chassis. One Achilles tendon and both knees are damaged. (Prompting a joint specialist to tell him not long ago that his running days were over. Of course, that was before Whitlock demolished the world marathon record.) He does have an unbelievable ticker, and so on this measure of cardiovascular health Whitlock is indeed a study. But in other ways, possibly from oxidative stress, Whitlock resembles Father Time himself.

For Olga, whether to train the Ed way or the Earl way wasn't really hers to choose: the body type she inherited made it an easy call. If she were built like Whitlock, she might have leaned toward distance and away from the short stuff and the throws. But, she says, "you work with what you're given."

The day after our CrossFit visit was not a good day for me. I couldn't even take off my sweater. Had some trouble getting my fork to my mouth.

Olga fared much better. She had a bit of soreness in the rear shoulder area—a few muscles she doesn't use much were awakened. That's paleo-grade resilience. True, she hadn't totally emptied the tank for Sheppy.

"Look, at this stage I don't have to prove anything to anyone," she said.

Aging Like an Astronaut

The times I think that Olga and I are a little alike—hey, we both work out!—I have to remember this very profound difference between us: I am sedentary and she is not.

I get my hour at the gym every other morning, fifty minutes of screaming hell on some aerobics machine. And then, pleasantly narcotized, I settle into a chair for hours, without moving a muscle, in a happy stupor. It looks like an evolutionary fitness plan modeled on the habits of Fred Flintstone.

This is a pretty typical pattern of "knowledge workers" in the present age. We are mostly very still, sitting in traffic or, when we get to where we're going, resuming our inertia in a work space so ergonomically efficient we barely have to twitch to field a phone call or brew a coffee. And then, in the evenings, we relocate our butts from dining room chair to TV room sofa. If urine bags strapped to our calves were just a little more comfortable, we'd probably go for them, too.

There's a perverse pride in developing the ability to strap in and blast through workdays like that. For six or eight straight hours you are driving the wheels of art—or commerce—with nary a lapse in concentration. Being comfortable helps. Why stand when you can sit, sit when you can lie? Didn't James Joyce write stretched out on a bed?

This all turns out to be a terrible idea, healthwise. Extreme sedentariness, if you want to call it that, is a scourge in ways we're only beginning to understand. If stress is the "black lung" of the information age, sedentariness is the equally menacing lifestyle by-product of modern times. Maybe it's our syphilis.

We evolved to handle heavy physical demands. So the body gets confused when no demands are made on it for long periods. It doesn't know you're just sitting around by choice; it figures you're in some kind of trouble—perhaps stuck somewhere under siege, without the strength even to send up a smoke signal. It mounts an emergency response—part of which is to shift metabolic resources. Instead of burning glucose it starts storing it as fat.

If you enforce inactivity in mice by holding their hind legs still, one study found, within twenty-four hours insulin regulation was messed up. Within forty-eight hours, those mice's hormonal profiles had completely changed. Even "fit" mice now showed the same insulin-resistance levels as couch-potato mice.

Now, inside those mice (and presumably inside knowledge workers as well), some dangerous mechanisms have been set in train. Glucose and insulin levels spike sharply after every meal. At the cellular level inflammation is happening. Down this road lie type-2 diabetes, cardiovascular disease, and as many as eighteen other chronic afflictions that are now sometimes dubbed "diseases of civilization." No wonder researchers have started calling inactivity "Sedentary Death Syndrome."

The Norwegian company Myontec has developed motion-tracking underpants with little sensors sewn into them that measure muscle activity and inactivity. In tests, subjects relaxed into those cyber-garments. The moment they stood up the graph looked like the jolt to life of a flatlined patient. When you unplug your muscles from the grid like that, the results are as predictable as if you stop watering your plants. Sarcopenia—muscular atrophy—ensues. "Each of us over age thirty-five is experiencing sarcopenia this minute," says McMaster kinesiologist Stuart Phillips. "We just don't realize it."

A few years back I discovered a new position to work in: semi-reclined, deep in a wingback chair, feet up on an ottoman, computer on lap. It simulated the business-class airline seat I once got upgraded to and in which got so much done I became convinced that complete comfort is the best way, and maybe the only way, to coax those shy deep thoughts from their burrows. It's especially good in winter, what with the pleasant warm feeling as the laptop battery toasts the skin on your thighs.

I mentioned my canny strategy recently to Joan Vernikos—the godmother of "inactivity studies." (She edited the first textbook on the subject, in 1986, called simply *Inactivity,* and in 2011 authored a book called *Sitting Kills, Moving Heals.*) There was a long silence. Then she laughed. Then she apologized for laughing. "Um," she said. "I think you're going in the wrong direction."

Vernikos is a physiologist and the former director of life sciences at NASA. The agency first hired her during the Gemini program to help astronauts deal with the stressors they face in space. But it didn't take her long to realize that about the most stressful time for astronauts came after they landed back on Earth. They were too wobbly to stand and they often blacked

out. Tests revealed physiological changes—among them a thin-
ning of muscle and bone. What was happening to astronauts in
space looked like human aging in overdrive.

It wasn't that they were "unfit"—indeed the astronauts had
been working out up there on stationary bikes. The missing
ingredient, Vernikos guessed, was gravity.

At the Ames Research Center in California, she began running simulations of reduced gravity effects with volunteer
subjects. She first immersed them in tanks of water, but soon
switched tack for ethical reasons (it turns out, unhandily, that
if you put people in water for more than six hours they start
showing symptoms of psychosis). Vernikos had volunteers lie
in bed for long periods, which dials down gravity's impact by
distributing it. And sure enough, subjects who signed up for
"extended bed rest"—a job that sounds cushier than it actually
is—routinely fainted when they stood up. Sensors in the neck
and chest that regulate blood pressure get messed up when we
don't give them feedback by changing posture many times a
day. If you're going to spend whole days lying down, Vernikos
found, you have to stand up at least thirty-two *times* a day to
keep those sensors in tune. Otherwise blood pressure can
become dangerously erratic and nutrients to muscles and bones
are interrupted. The heart no longer fuel-injects blood from the
legs to the head the moment that task is required. It is almost as
if a whole raft of body systems forget their jobs.

Vernikos ran a second study with subjects who were merely
sitting for long periods, rather than lying—she turned gravity
up a bit, you might say. The same effects appeared, "although
they took a little longer to develop."

And then the other shoe dropped. While visiting her father
in a nursing home one day, Vernikos felt a déjà vu. The old
folks' symptoms were familiar. They would get out of bed "and

have trouble with their balance and their coordination. And their muscles were weaker and their bones were weaker—the same things we had observed with the astronauts." That's when it occurred to her to ask: "What if we've been looking at the question backward?" The symptoms of being in space, everyone agreed, mimicked age-related decline. But what if the age-related decline she was witnessing mimicked the symptoms of being in space? If that was true, mightn't those symptoms to some degree be reversible, just as with the astronauts and her volunteers? What if these people in the nursing home just needed a little more gravity in their life?

Vernikos started refining a philosophy that she preaches to this day. We need our muscles—really our whole bodies—to be challenged by resistance, and gravity is the best agent of resistance we have.

"If you take plants and put them on their sides they grow upward, even in the dark, where they're not getting a cue from the sun," she says. "They use gravity. And so do we—or at least we should. It's how we orient ourselves around this force of gravity that stimulates every cell in our body, and provides the stimulation we need to be alive on earth. It's how we evolved."

On Earth, where gravity is free and available, it makes sense to concentrate its force.

Standing up a lot is "the single most important habit we can acquire," believes Vernikos. It is the original paleo workout: rising from the squat position to, say, throw another log on the fire. "It's tuning your body, using gravity, so that when you need to chase the prey or climb the tree in a hurry, you can do it," she says. "It's not just endurance—it's the sprinting effect. It's keeping the system alert so that it can respond."

The second-best exercise, bang for buck, is to stay standing. True, it doesn't burn many calories, but in terms of muscle activation, standing for two hours has been likened to going for a two-mile run—which may be why museum-going is so perplexingly exhausting.

Social scientists who study the world's "Blue Zones"— pockets where extreme longevity is common—note that the residents of those places are rarely "exercisers," per se. They don't pursue "fitness," or even think of themselves as fit. But fit they are, because their environments happen to be constructed in a way that forces them to move. In those towns and villages people climb ladders or stairs to their homes, walk to water wells and markets, or trudge up and down hillsides *every single day.* It is "work" that's not work at all, because it so seamlessly, invisibly, becomes integrated into life.

This is not our way in the modern West. Our way, for the most part, is to sit staring at screens. When people do that, it is becoming abundantly clear, they develop health risks that can't be fixed by periodic bouts of even intense exercise.

The idea that fitness is something that can be packaged into forty-five-minute bundles taken like a superpill a few times a week, thereby fulfilling your exercise allotment, may be a kind of mass delusion that only a minority of wise souls like Olga have escaped. It's wrong in a lot of ways that seem right.

Getting a blast of cardio every other day isn't enough to be "fit," although—and this is the problem—many of us think it is. That sense of virtuous accomplishment a workout gives you can backfire, it seems, by making you coast and cheat in the rest of your life—emboldened by what one Chinese research team called a sense of "illusory invulnerability." In the Chinese study, subjects who thought they'd taken a dietary supplement actually made less healthy choices the rest of the time, such as walking less. By getting my workout in, I feel accomplished and actually do less than if I hadn't worked out at all but rather built movement into my daily routine.

The cannabis-y buzz we get after a hard run basically wipes out our fidgety restlessness: so nice. But fidgety restlessness is what makes us adjust our position continually. And moving, even a little bit, continually is what it's all about. James Levine, an inactivity researcher at the Mayo Clinic in Rochester, Minnesota, found that his test subjects who "unconsciously moved around" didn't gain weight; microfidgeting was like a workout spread over a whole day.

The difference in health benefits between sitting stock-still and walking across the room to get a pen, Swedish researchers recently concluded, is probably greater than the difference in health benefits between being an "exerciser" and a "nonexerciser." In other words, if I found reasons to get up and move around regularly all day, my health would improve *even if I never got another run in again.* Ideally, you want to do both: exercise *and* move around. Which is, of course, what Olga does.

If people of Olga's generation were and are healthier, it's partly because of simple good habits, Vernikos believes.

"I ask people today, Do you get dressed standing up? You'd be amazed how many don't. Young people. *They sit on the bed to put their pants on.* Or their shoes on. Older people are much more likely to get dressed and undressed standing up. It became a habit for them before the culture decided that we should do everything sitting down.

"I would bet you that Olga, if she sees a thread on the carpet, bends down and picks it up."

(She does. Also coins. Even pennies.)

So in a weird way, you could say—and Vernikos *does* say—a bigger part of the secret of Olga's robustness may be what she has done off the track, not on it.

Forget aerobic versus anaerobic and all those other distinctions for a second; this is much more basic. Olga has just rarely been still. From the farm to the schoolhouse to night school to the track, she has been going somewhere or heading back, doing some job or cleaning up after it. Even back home in the confines of the basement suite, she's likely to be out back in the vegetables, or making a complicated spaghetti sauce, or gathering material for her memoirs, or going up or down the stairs (a trip she makes, by her own estimation, "probably fifty times a day").

Think of gardening: you stand, you crouch, you stand, you crouch, and then you carry the now heavy weed bucket to the compost. (Only this year did Olga stop tending a garden the size of the whole backyard; now she's downsized to just a small plot near the door.) Even puttering in the basement suite, she is staving off sarcopenia and keeping countless body systems maintained. This is Olga's routine.

Now compare mine. If a chicken moved as little as I do over twenty-four hours, the farmer would be legally prevented from calling it "free range."

Both Olga and I exercise, but she moves when she's not exercising, and I don't. And *that*, Vernikos says, is the rub. Olga is older than I am. But 95 percent of the time, I am *getting older* faster than she is.

Here we must acknowledge an apparent contradiction. The work of Vernikos, set against Scott Trappe—who as a postdoctoral student conducted experiments alongside her at the Ames Research Center, and who would go on to become a prominent exercise physiologist in his own right—shows how these questions surrounding optimal health and longevity are still in flux. Vernikos and Trappe agree to a point: exercise is hugely important. Resistance exercise in particular is crucial. Sitting like a lump for long stretches of time is a travesty.

But the two researchers part ways on the question of how hard to push. Vernikos believes that job one is just to stand and move, continually: no muscle-shredding gym workouts needed.

Trappe holds that, in terms of physiological payoff, "light activity" can't hold a candle to breaking a sweat. When Trappe studied the exercise programs of nine astronauts on the International Space Station, he concluded that the reason they weren't able to stem the effects of "aging" up there is their regime wasn't nearly hard-core enough. His prescription for NASA: heavier loads and explosive movements. "It's pretty clear that intensity wins up there," he told me. "And I would predict this to be the case as we age."

Both Vernikos and Trappe could cite studies in their defense. Can they both be right? They can, in the sense that the correct approach depends on what you're mostly after: general health or performance. As we age the former becomes more important for most of us. But those select few who are still thinking about high performance as they get very old are a godsend to scientists—as we're about to see.

6

Tests of Body

OCTOBER 2010. IN a stuffy room on the third floor of the Montreal Chest Institute, Olga stands before a treadmill. She wears a plastic crown that holds in place a snorkel-like breathing tube. She has donned a stretch vest fitted with electrodes that will measure the changes in the electric current her heart is producing. "Do you feel like Lady Gaga?" asks Dr. Tanja Taivassalo. Olga allows, with a shy smile, that she does.

Taivassalo is the McGill University cell-mitochondria specialist who had first contacted Olga, back around the time Olga and I first met, and started laying the plans that would lead to this day—the first step in an ambitious series of tests designed to assess the physiology of elite masters athletes. Whatever is happening with Olga amounts to a spark of hope for the rest of us. Understanding what makes her go could yield practical benefits not just for the "old old" like her, but for the younger old, who could perhaps better slow the rate of muscular and aerobic decline. (Not to mention the degradation of balance, which leads to falls, which are the beginning of the end for so many seniors.)

In this pilot study, Olga, alone in her Lady Gaga vest, is the guinea pig. Her results will be leveraged for more funding to expand the study to include elite masters athletes from all over the world. And as much as Taivassalo wants new blood, she also wants Olga back in two years, to see what has changed in her body in that time. She is shining a pinprick of light into the physiological void. "No one, as far as we know, has ever done these measurements in someone this old," Taivassalo says.

Today's treadmill test will yield a VO_2 max score for her aerobic capacity. Olga's job is to go for it, against increasing resistance, as long and hard as she can. A grad student rubs Olga's right ear with a cream to draw blood to the surface for extraction. Olga is now hot and her throat is dry and her ear hurts and the mouthpiece chafes. She is ready to rock.

There's a certain amount of locker-room hup-hup. Guesses are exchanged on how high Olga's heart rate will climb. (The old rule of thumb was that you should never rev your heart beyond 220 minus your age. That would suggest a maximum, for Olga, of 129 beats per minute. But recently a more complicated formula was developed to boost accuracy in those higher registers; it allows that the older heart can handle more than we once thought.)

There is mild concern that Olga's stubborn stoicism in the face of pain will make her push herself to dangerous extremes. The Finns have a word for this trait: *sisu*. It's something Taivassalo is familiar with: her Finnish, marathon-running dad has it, too. And so a pulmonologist, "Dr. Ben," is on hand. His job is to watch Olga's EKG profile for abnormalities, and to call a halt if he sees any.

The treadmill hums to life.

Thirteen people are crowded into the room. Taivassalo tells Olga to tune out everyone except her and Dr. Ben.

Russ Hepple stands close by, keeping his eye on a digital number on the console—the ratio of carbon dioxide to oxygen in every breath. (The ratio climbs as the going gets tougher.

When the first number exceeds the second, that's about when most people think of quitting.)

Hepple is studying ways to tailor exercise regimes that will maximize the aging body's adaptations. Taivassalo met him at a physiology of aging conference. She was so impressed, she married him. Together, the couple seems poised to establish a formidable convergence of research—"he on the animal side and I on the human side," she says, slyly.

Olga's heart rate soon climbs beyond 110. It hits 120, 130. The treadmill steepens to a 10 percent incline. Olga is sweating. The room feels like a foundry.

Very quickly Olga's perception of discomfort, on a scale of 1 to 10, spikes from 3 to 9, then 10, and the test is suddenly done. She has trudged on the treadmill for seventeen minutes. Her heart rate reached 138. Her VO_2 max was around 16 milliliters per kilogram per minute—an unremarkable score for a young woman, but very good for someone in their 90s.

"It was a perfect test," Taivassalo says. "Nothing broke—that doesn't always happen."

"I said an extra prayer, that's why," says Olga, toweling off.

"The Most Beautiful Sample Ever"

Two days later, Olga lies on an examining table in the storied Montreal Neurological Institute and Hospital, where Wilder Penfield mapped the human brain nearly a century ago. She is about to make her most important contribution of the visit.

"Contract your thigh muscle, please," Dr. José Morais says. "Whoa." The muscle shrugs up visibly as she tenses. Russ Hepple and Tanja Taivassalo have borrowed Dr. Morais from the geriatric medicine department to perform a biopsy. He looks for a suitable spot for a lidocaine injection. The capillaries are typically more fragile in older people than younger people, so Dr. Morais is used to taking care at this stage. But most older legs don't look like this.

For Taivassalo and Hepple, it is a precious opportunity. A scientist can strength-test till the cows come home, but a bit of actual muscle to put under a microscope is invaluable. Muscle contains what Hepple calls "biomarkers of aging"—evidence of changes over time in its structure, biochemistry, protein expression. These mark the body's decreasing ability to withstand the stresses it encounters—"some from outside us, like infections, and some from inside us," such as the cellular trash that builds up through normal body functions such as breathing and metabolism. "In essence, they tell us how well Olga has handled the very things that cause most of us to age and die at or around age eighty," Hepple says.

Taivassalo and Hepple are each eager for a bit of tissue, but for slightly different reasons. Taivassalo is curious: Just how healthy is that muscle of Olga's? How damaged by age is it (something you can guess by the number of free radicals)? How well is it able to protect itself (the number of antioxidants)? How good is it at producing energy (the number of mitochondria)?

Hepple wants some muscle fibers to test a theory about the role of neighboring nerve cells in aging muscle.

With a scalpel, Dr. Morais makes a little incision, and then inserts a gleaming silver instrument that looks a bit like a wine corker trailing a suction tube. Suction pulls the muscle sample into the cannula where a little blade chops it off.

The first sample is not very good: a little blob of fatty tissue. Hepple and Taivassalo try not to react, but their slumped shoulders betray their disappointment.

The doctor goes in again, deeper into the same hole. It is a fairly invasive procedure; and because older muscle takes longer to heal than young muscle, the timing of the tests would become another logistical factor to juggle as the database of elite athletes expands. It can't be done close to an upcoming competition.

This time the sample is everything the scientists had hoped for. Where old muscle is grayish, this little ort is as red as steak tartare.

"Oh, that's a beautiful piece!" Hepple says. He tweezes it into a petri dish—his share. Then he disappears upstairs to flash-freeze it in liquid nitrogen.

Dr. Morais harvests a third bit of flesh. This one is even juicier, and long, like a small caterpillar.

Pronouncing it "the most beautiful sample ever," Taivassalo puts it on ice in a Styrofoam container, like sushi takeout, and then she and her grad student disappear out the door with it. The grad student will speed it to a lab at the University of Montreal, there to tease out individual muscle fibers and put them in a respirometer. In the parking lot Taivassalo leans close to her student and tells him to guard that sample with his life—as if it were the roe of the Salmon of Knowledge.

But if Olga's muscle is being treated as a precious jewel, everyone is quick to reiterate that it is incomplete by itself. A data point of one isn't terribly useful, scientifically. (The scientists at the Beckman Institute were, of course, up against the same problem. Olga was fifteen years older than anyone else in their database. "If you ask why Olga is the way she is," the psychologist Justin Rhodes told me, "it's going to be a combination of genes and environment. If you want to know what those genes are, you'll never be able to find it. There's no way statistically. She's just one person.") Call it the conundrum of the extreme outlier. She frustrates our attempts to understand her because she is unique and therefore beyond direct comparison.

You could say that the science of aging depends on the hunt for more Olgas.

Which is why, as per the plan, Olga finds herself back in Montreal two years later for more productive punishment. This time she has company.

Widening the Circle

Olga is climbing out of a densitometer when I arrive at Taivassalo's lab. It looks like a tanning bed and reads bone mineralization and body composition with low-radiation X-rays. A beautiful set of color images issues from the printer. It is the Illustrated Olga: skeleton, muscle, and fat.

"Slightly higher fat content than I would have expected," Taivassalo muses, glancing at Olga's results. "But you know, it probably helps her."

Olga carries around 130 pounds on her five-foot frame. A lot of glossy magazines—the kind that equate health with dieting and success with celebrity—would say that's at least twenty pounds too much. But those magazines are wrong.

The evidence around body mass indexes and mortality rates is clear: if you're not prepared to semi-starve yourself, the best strategy for a long life is to be a shade zaftig—*as long as you're fit*. Studies consistently show body mass indexes in the mid-range in cardiovascularly fit people are most strongly linked to health over the long term. As she does in so many metrics, Olga hits the sweet spot here, with a BMI of a little under 25. That's in the "normal" range (18–25), but just.

Back at Taivassalo's lab, three of this week's new test subjects are relaxing after their cardiac tests. Two of them are British masters runners whom Taivassalo recently recruited at the 2012 World Masters Athletics Indoor Championships in Jyväskylä,

Finland. Colin Field is a barrel-chested sprinter. Arthur Kimber is a slight, soft-spoken middle-distance man whose claim to running fame is edging his boyhood idol, former world champion Chris Chataway, in a 1,500-meter race at age 60.

The third is a petite woman in jeans and tennis shoes. When I first see her from the back, I think she might be a teenager. She is Christa Bortignon, who is here because of Olga, and who has a decent chance of becoming Olga 2.0.

Four years ago Bortignon read a newspaper article about Olga, and because they were practically neighbors, she called her up with congratulations and questions. Who can enter these meets? Where are they held? How many events can you do?

"Meet me at the track in half an hour," Olga replied.

Around the oval they walked together. Bortignon had been a nationally ranked tennis player before arthritis took her out of that game, and now, at age 72, she was hungry for a new outlet.

"There wasn't very much mentoring to be done, even from the beginning," Olga recalls. Bortignon had blazing speed and would soon establish herself as the world's fastest 75-year-old woman. So meteoric was her rise that she aroused suspicion and was drug-tested twice.

In the two years since those first tests on Olga, the McGill study has begun to grow as the scientists hoped. The list of world champions over age 75 who have come to McGill for testing includes miler Jeanne Daprano and middle-distance man Earl Fee, marathoner Ed Whitlock, and sprinters Bortignon and Bob Lida. The project is starting to feel historic, like that famous Art Kane photograph of fifty-seven jazz icons somehow roused from sleep and mustered on a Harlem street in a never-to-be-repeated convergence. (True, none of the other athletes have been women in their 90s—so Olga has still not received the apples-to-apples treatment, scientifically speaking. But word of the study is spreading, and Hepple and Taivassalo have high hopes for eventual inclusion from Olga's closest proxies, wherever on the planet they might be.)

It has been a grueling week. Scientists checked the pumping power of the athletes' hearts and the efficiency of their lungs. In blood samples, they hunted for inflammation markers, whose ratio gives a hint at how the immune system is functioning, and checked hormone balance, which is thought to be linked to aging.

A main focus, as ever, was to sleuth out the mystery of why muscle degenerates with age. Is it a problem within the cells themselves or something else about the body they're floating in?

This has become a hot question ever since a Stanford stem cell researcher named Saul Villeda discovered a kind of "vampire effect": old mice that received blood transfusions from young mice suddenly started performing better on memory tests. And, conversely, young mice that received blood transfusions from old mice immediately started performing worse on those tests. The difference seems to be the "soil"—the blood "serum." Blood serum is full of things—circulating proteins, growth factors, and immune-system boosters—that age tends to deplete.

If the blood of trained masters athletes has some of the properties of young blood—if it is "rich soil"—then that could help explain why Olga's muscle stays so strong. To test the theory, stem cells would be harvested from couch-potato controls of Olga's age. They'd be dropped into a petri dish full of Olga's blood serum. If those "old cells" started growing in the "young blood," then "that's good news," says Sally Spendiff, a British postdoc student working with the Taivassalo team. "It suggests there's nothing intrinsically wrong with the cells—theoretically they can be restored." The trick would then be to figure out what the good stuff in the superpremium blood actually is. The endgame could be a synthetic soil chock-full of those healthful proteins. Elderly sedentary folks receiving injections of it would, in theory, become a little more Olga-like.

In this round of testing, Hepple and Taivassalo were applying a full-court sensory press on Olga's muscle: looking at it, watching it work, and even listening to it. Yes, if you crank up the volume on the electrical impulses in the muscle cells, muscle does make a sound.

In Taivassalo's lab, Olga is met by Matti Allen and Geoff Power, two young researchers from the University of Western Ontario. A popular theory about why athletic performance dives as we grow old is that the heart loses strength, and so the muscles don't get enough oxygen. This is the conjecture that recently sent a team of Mayo Clinic scientists to the thin air of Mount Everest's base camp to study how reduced oxygen seems to steal muscle mass, just as aging does.

But the UWO group, led by Charles Rice and Tim Doherty, have a different theory, which jibes with what Hepple has been guessing all along. They think it involves "motor units"— parts of the nerve that stimulate individual muscle cells to contract.

As we age, these motor units die off. Sitting there in anticipation, hearing nothing from the nerves, the muscle cell stops contracting, and eventually it dies, too. (That would make muscle loss, like so much else that's wrong with the world, a communication problem.) The good news is, when you exercise a muscle you retain the motor units that plug in to them. Rice and Doherty wondered how deep into old age the neuroprotective effect of exercise works. They knew it worked at least into our 60s—in a previous test, runners that age had plenty of good motor units in their leg muscle still. But nobody had tested athletes older than that. To what degree did "use it or lose it" apply to them?

Olga settles into a leg-extension machine, the kind you see in gyms. She raises her foot, flexing the muscle around the shin. A crackling issues from the speaker: a chorus of neurons firing. It's like picking up chatter from a phantom world—a world you have to have faith, or at least serious amplification, to appreciate. Though nobody knows it in the moment—it will take months for her muscle to be analyzed—Olga's muscle is producing a sound twice as complex as most women her age would make with theirs, because roughly twice as many of Olga's motor units are active. If the front tibia muscle of most 90-year-old women is a small city orchestra, Olga's is the New York Philharmonic.

Next comes the watching. Olga heads for the basement of Montreal General Hospital for a CAT scan.

As we age, and our muscle starts vanishing, it gets replaced by fat. If we were cattle, we would become tastier, our meat more marbled. In cross section on the monitor, steaks are exactly what Olga's thighs look like. A thick ring of subcutaneous fat surrounds them. But inside that ring, hold on: the "meat" is still startlingly red and lean. This looks like awfully young muscle. Since her last visit for testing, Olga has tapered way off her weight-training regimen, so her legs ought not to look like this. Whatever was protecting her muscle then is likely still in play now.

The final proof will be in the pudding—how strong she still is.

Taivassalo gathers her athletes in the weight room and commandeers a bench-press station. Students in giant basketball shorts move about, carrying plates for their weight stacks. Colin Field, the 80-year-old British sprinter, lies down and blasts off a long set with 175 pounds on the bar. Then, still restless, he slips away to a quiet corner of the gym and starts doing biceps curls to pass the time before his next test. It's Olga's turn.

She lies flat, takes a deep breath, and proceeds to bench press sixty pounds fifteen times. Taivassalo checks her notes. It is marginally *more* than she squeezed out two years ago. How could that be?

During that pilot study, when Olga first settled under the bar here to test her strength, she drew a crowd. Today, the students are oblivious. A couple of young guys agree to spot Olga, but then get distracted by some girls. Nobody seems to appreciate that the little old lady over there bench-pressing a not-insignificant amount of weight is close to 95 years old. It's not something they are likely to see again.

On the last night of testing, as a reward to the athletes for their week's hard work, Taivassalo and Hepple take their test subjects out to dinner at a downtown brasserie. Hepple looks around the table. Athletes tuck into swordfish and horse steak. (This is French Canada, after all.) There are a couple of 80-year-old road cyclists, two 80-ish runners from Britain, and Christa and Olga.

Exercise is the thing they all most obviously share, apart from their scandalous youthfulness. It's almost irresistible to make the causal leap. But what if we have the equation backward with people like Olga? Perhaps it's not that she's aging slowly because she's a champion athlete, but rather she's a champion athlete because she's aging slowly.

Hepple has as nuanced an understanding of the benefits of exercise as just about anybody on the planet. On one hand, he's a little blown away by what we now know it can do, how comprehensive are its effects on tissues throughout the body. But at the same time he's aware of where the evidence stops. And beyond that line we are simply overselling exercise as the elixir of extended life.

"Look, exercise is great," Hepple tells me across the table. "It's better than any drug ever invented. It goes so wide. *But.* As someone who has worked this area for his whole career, I can tell you, it's not the be-all and end-all.

"What these people have," Hepple says, "it's something special. Something else is going on. And I want to know what it is."

The Body's Early Warning System

When masters athletes talk about toughness and resilience, they don't mean quite the same thing scientists do. They don't use terms such as "corticoid receptors" or "allostatic load." They mean *the ability to keep showing up for meets.*

The aging athlete, you might say, is cruising down the freeway over a sketchy part of town. The off-ramps fall away into darkness, and all the on-ramps are closed. The moment something fails and he or she has to pull off that road, life changes.

One of the great mysteries of Olga is how she competes, year in and year out, without getting hurt. Her occasional injuries—biceps, quad, shoulder—are never too serious and she heals fast. And she's barely been sick even a day in the past fifteen years.

"Her body is seemingly bulletproof," says Ken Stone of masterstrack.com. "She'll come in day after day after day at a multiday track meet and excel at everything she does. If you ask a forty- or fifty-year-old to come and do the same events, they blow a gasket. But I have never heard of Olga pulling out of an event because 'I hurt myself yesterday.'"

Why does Olga's ridiculous eleven-event load not produce injuries?

Resilience, as we've seen, is partly about how the body learns to manage stress. But resilience is also about heading off insults to the body before they happen.

At the recent championships in Finland, "Christa mind-boggled me into signing up for the pentathlon," Olga recalls.

But in the middle of the night before the event, Olga sat up in bed. "I thought: what am I doing?" The pentathlon requires an 800-meter sprint. And hurdles. "I'm afraid of hurdles," Olga says. "This is sort of a phobia." Her subconscious had raised a big red flag. "I really don't need to do this, do I?" she thought.

A strong self-preservation instinct courses within Olga—the same one that prevented her from stepping out onto the busy street in that virtual-reality environment in Illinois. I think of it as a kind of early warning system.

Many masters athletes pop anti-inflammatories like candy. Olga doesn't. Anti-inflammatories mask pain so that athletes feel they can then push on, often into the damage zone. But if you're alive to the first stirrings that there's a problem—it may be a problem of technique, or a muscle imbalance, or just overuse—you can step away before you start shaving away the cartilage, creating more inflammation, compounding the problem, and eventually hobbling home on your new titanium knee (or rolling home with your new titanium hip). By stepping away and resting, she leaves the repair work to nature, which has had a few million years to get it right. "The body," she says frequently, "is built to heal itself."

When something starts to feel a bit off, Olga stops doing it. When she senses she's being overtaxed, she withdraws from the tipping-point event. She doesn't *owe* anybody anything— not some sponsor, not meet organizers, not fans. The buck stops with her. Her position reminds me of that of Kenyan Patrick Makau Musyoki, the world-record holder in the marathon (2:03:38). Like Olga, Makau is coachless. That way, "you can listen to your own body and take time to recover after training," he told a reporter recently. "Sometimes a coach pushes too much."

There is a name for this faculty of superattunement for what's going on inside your body. It's called interoception. To be interoceptive is like being introspective, but to sensations rather than emotions or thoughts.

Receptors throughout the body are continuously recording information. They track changes in cardiac output, hormonal and metabolic activity, even immune-system activation—the level to which we are "run-down." There are receptors scattered all through the fascia—the network of connective tissue that wraps every organ and every muscle and connects to every ligament—and they are hair-trigger transmitters of "tensional strain" anywhere in the body.

All this information goes into the brain. And here differences between people emerge. Some of us are better than others at picking up those faint signals and dredging them into consciousness.

There are signs that Olga is adept at this. She says she can feel a cold coming on, and when she does she takes a baby aspirin to "get out in front of it." She's conscious of what to eat, and how much, and when. She has a great line of communication with her gut.

Killer attunement to your body's signals is a fine thing to have if you're an athlete—and not just because you can stop before you cause an injury. Studies have found that it actually boosts performance. Elite marathoners pay close attention to their internal landscape—breathing, heart rate, cadence, how the muscles feel—while nonelite marathoners do just the opposite: they try to distract themselves from the pain. Some evidence suggests that tuning in, rather than dropping out, could have other benefits, including healthy longevity.

Can interoception be measured?

Turns out, there *is* a simple test that seems to serve as a rough gauge of this faculty. It is the ability to tell time without a watch.

University of Arizona neuroanatomist Bud Craig figured out

the link. Craig noticed that two seemingly different systems—
the way we experience time and the way we pick up visceral
and emotional signals—share neural pathways. The same parts
of the brain "light up" when we're estimating the passage of
time as when we're estimating pain or decoding emotions. All
of the data we get, from both inside and outside the body, is
converted into a series of what Craig calls "global emotional
moments." Those moments are discrete little pulses, and they
amount to an accurate internal timekeeping mechanism. All
that remains is to convert "body time" to clock time. And we
do it: instantaneously and unconsciously. Sensitive people do it
very, very well.

On the airplane to Chicago, I had given Olga the body-
clock test.

We'd been aloft for a while, and she was deep into her
Sudoku. I watched her get the little dopamine hit from finishing one and being propelled to the next one in high spirits.
(Meanwhile, I was fully taxed just trying to untangle my iPod
earbuds.)

I brought up the stopwatch function on my phone and set
up the experiment.

"I want you to tell me when a minute has elapsed. And I'm
going to distract you, so you don't count. I want you to kind of
just set the alarm inside your head for sixty seconds and then
tell me when it goes off."

Go.

"Tell me a little about what your grandchildren are up to—
Matthew and Alesa."

"Well, Matthew's traveling all the time—he's in Colombia
now," Olga said, and she started filling me in, and there was
pride in her voice, and I knew she wasn't counting. Suddenly
she said "Stop."

"Good. Let's try it again."

We repeated the experiment four times. She was almost dead-on consistent, within two seconds of the same mark each time. But here was the thing: she was always short. It was as if her internal clock was Swiss accurate, just calibrated to a forty-eight-second minute. Almost as if time had been devalued by 20 percent.

"That's unusual," I told her. "Most people find time flies as they get older."

"Time used to fly," she said, "but in the last few months I've noticed it slowing down." These days, she will prepare to go out to an appointment that's an hour away, "and I'm ready fifteen minutes early. Why I don't know."

The body-clock test is in some ways more than a measure of mere timekeeping; it is a measure of sensitivity more generally. If you're alive to the fine calibrations of the signals from your own biochemistry, you're probably pretty intuitive about the signals coming from others.

Once, in an airport, we passed a teenage girl sitting on the stairs, her face in her hand, crying. Olga instinctively veered toward her. I think Olga feels the way dogs hear—on a broader spectrum than most people. I know she feels more deeply than I do. The Buddhists would say she is powerfully self-aware, and self-awareness is the way out of "dukkha"—suffering and chronic disease. It is the path to good health and long life.

7

Habits

It's time to expand the scope of inquiry. After all, when even heavy exercisers are doing, well, *something else* 95 percent of the time, it stands to reason that important clues are hidden in our quotidian routines. How might Olga's other habits be contributing to her longevity?

Sleep

Almost every night Olga wakes up between two and three a.m. She'll lie there in bed. If she has come out of REM sleep and there's a fresh dream lingering (a rare event), she may savor it as its memory slowly melts away. And then she will do one of two things: let herself go back to sleep or truly wake herself up.

On the nights she wakes herself up—every other one—she starts in with her "OK" exercise program, a meticulous routine she invented—and named to riff on her initials—that she has been following faithfully, in the quiet of the night, for a dozen years. In a meditative state, her mind idling, she systematically kneads her skin and stretches her muscles. Ninety minutes later, well worked over, she goes to sleep again.

This pattern of activity has a name. It's often called "segmented" or "polyphasic" sleep. And there's good evidence that, for 99 percent of our history, it was how we passed the night.

Roger Ekirch, a professor of early American history at Virginia Tech, studied the sleep patterns of preindustrial people for his exhaustive 2001 book *At Day's Close: Night in Times Past.* In prehistoric times it was dangerous to go offline all night, what with prowling predators and wandering-eye neighbors—and anyway, darkness lasted longer than you needed to sleep. So you sectioned the night up. You might drop off a couple of hours after sundown and sleep for four hours (the "first sleep"), then be awake for an hour or so—talking or snacking or praying—before submerging again for a "second sleep." That was more or less the norm right up until the nineteenth century, when electric light extended the day and imposed a new pattern on our nights and days—sixteen hours on, eight hours off.

Is Olga-sleep actually healthier? We do know it produces a different body chemistry.

Around five years ago, Thomas Wehr, a psychiatrist at the National Institute of Mental Health, best known as the codiscoverer of seasonal affective disorder (SAD), conducted experiments to simulate prehistoric sleep. Wehr turned the sun out on his subjects for fourteen hours a night, for a month. They fell back into preindustrial rhythms; after sleeping for four hours or so they surfaced into Olga-like "non-anxious wakefulness" for a couple of hours before slumbering again for another four hours. In that wakeful interval they observed, among other things, a spike in the hormone prolactin, which is involved in the immune system and believed by some to be one of our first lines of cancer defense.

The subjects were waking up a little but not all the way, in the nighttime interregnum. The lack of light was key. If you awaken in the middle of the night and snap on the bedside lamp, it signals "daytime" and interrupts the flow of melatonin—a hormone that not only helps regulate our circadian rhythms but has been implicated in the aging process itself. (Melatonin disruption has now been so credibly linked to elevated cancer rates that the World Health Organization has labeled shift work a "probable carcinogen.") In perhaps a stroke of accidental genius, Olga got this exactly right. Her mind is awake and idling, and her pineal gland keeps producing melatonin while she does her meditative business in the dark.

A good night's sleep "charges the batteries," we are told. The energy you use today was built last night while you slept.

Unfortunately, a lot of us find our sleep permanently compromised from about teenagehood on—first for play, then for work, then for the impossible work/family two-step, and finally for physiological reasons no one yet fully understands.

Ongoing sleep debt can actually do permanent damage. It disrupts hormones governing appetite—thereby "deceiving the body into believing it's hungry when it is not," notes David Agus, a professor of medicine and engineering at the University of Southern California. This can cause insulin to spike, which triggers an inflammation response. Sleeplessness appears to impair the formation of nervous-system tissue, and hamper the repair of cells damaged during the day. It's increasingly being seen as a stealth culprit in far more afflictions than it's officially blamed for. Some sleep researchers, including William Dement, founder of the Sleep Research Center at Stanford and the team that effectively "discovered" REM sleep in the 1950s, thinks a healthy sleep life is as strong a predictor of longevity as even exercise or heredity. (The sleep-longevity case is circumstantial; it's based on the increased incidence of health risks when sound sleep is taken away.)

For those who don't sleep well at night, there are ways to scavenge energy in the daytime.

Nerina Ramlakhan, a former research physiologist and now

a sleep consultant at a private psychiatric hospital in London and author of the book *Tired but Wired*, believes the secret is to honor our natural body rhythms.

We oscillate, 24/7, through "on" and "off" modes—a phenomenon dubbed the basic rest-activity cycle (BRAC), by the man who discovered it, University of Chicago physiologist Nathaniel Kleitman (who lived, incidentally, to be 104). The principle is this: Every ninety minutes or so the brain wants to lighten up; it wants to recover from the work it just did. At night we honor those ninety-minute cycles unconsciously, rising and falling on that tide, from shallow to deep sleep and back again. But during the day, when we're in control, we often plow through those signals. Ideally, every hour and a half we'd push back from our desks, flop down on the floor, and nap for fifteen minutes. That's a recipe for good health and maybe even long life. It's also a recipe, in most offices, for exploring other employment options. Ramlakhan counsels her clients to build in other kinds of rest breaks—be they "active" rests, such as a dreamy walk, or "passive" rests, such as meditation or prayer. Both can boost energy levels.

I've observed Olga's daytime routine to see if she follows that advice. She doesn't get much passive rest. But she does routinely get "active rest," mostly in the form of the Sudoku book she reflexively reaches for whenever a natural break presents itself. Sudoku is "rest" inasmuch as it interrupts the continuous incoming flow of new data, says Ramlakhan. The brain is working when you do a Sudoku, but it's not storing things—it's using RAM but not the hard drive. And that is regenerative.

The jury is still out on polyphasic sleep systems like Olga's. It's hard to know if the effect is idiosyncratic. Olga-sleep works for Olga, for her particular hormonal signature. Her body, perhaps heeding the call of some faint ancestral chime, wakes her up in the night and nudges her into periods of "active rest" during the day.

Still, Olga has other habits we can copy if we want to sleep soundly, in order to live longer and better.

She isn't torqued up on stimulants or refined sugar.

And, of course, there's all that exercise, which we can't ignore when examining her sleep habits. Exercise may be the best sleep aid there is. Physical exercise during the day has been linked quite precisely to faster sleep onset (in one group of children studied, "every hour of sedentary activity during the day resulted in an additional three minutes in the time it took to fall asleep at night"). Some data, including a recent study out of the Feinberg School of Medicine at Northwestern University, suggest aerobic exercise produces the fastest onset of the highest quality of sleep for people of boomer age on up.

But there's an inverse to this rule—poor sleep means poor exercise. You can't get the full benefit of a workout when you're gassed going in; at least 10 percent of cardio capacity is lost, studies show. Time-crunched, many of us face the Sophie's Choice of sacrificing sleep or exercise. If you're addicted, it's no choice at all: you exercise. Most physicians would say that's the wrong choice. And a coach who understands physiology would point out that it's counterproductive.

How much does Olga's athletic performance owe to good sleep? Cheri Mah, a researcher at the Stanford Sleep Disorders Clinic and Research Laboratory, answered that question indirectly not long ago.

Mah wondered what would happen if Ivy League college athletes—whose academic and athletic demands make them poster children for sleep deprivation—actually got the sleep their bodies craved. She recruited players from Stanford's basketball squad and put them on a regime of ten hours' sleep a night. Every performance metric improved—sprint speed, reaction time, shooting accuracy.

As heavy an exercise load as Olga's demands more than a

bare minimum of sleep, "to produce the steroid hormones needed to repair muscle and restore the integrity of the immune system," as Ramlakhan puts it. A few years ago, when she was hired to work with the Chelsea Football Club, Ramlakhan found a lot of the players weren't sleeping nearly enough for the volume of training they were doing. If she couldn't change their lifestyle habits—which often included playing video games deep into the night—she'd encourage them to snooze during the day. "I'd try to get them to have naps in the afternoon, even if just twenty minutes," she says. More sleep means more energy, better recuperation, and sharper preparation for tomorrow—however many siestas it takes to get there.

The "OK" Technique

A green, narrow-barreled wine bottle sits on Olga's bedside table, "so it's easy for me to reach over and grab it" in the night. Olga hits the bottle three times a week. She thinks this explains a lot. And it may well—but not the way you're thinking.

The bottle is empty. It's simply a massage aide.

To begin, she slides the bottle under her back. She feels the cool glass on her skin next to her shoulder blade. Her own body weight creates the therapeutic pressure; it's like a strong young masseuse from the old country is leaning on her. She rolls over the bottle, moving it down the muscles on one side of the spine, then repeats the process on the other side.

Olga ditches the bottle and then, following a few deep, tidal breaths, starts in with digital manipulation. She pinches the tips of her fingers and starts moving up into the hand. She works the skin like a baker. She moves up to the arm, the shoulder, the top of the head. She massages her scalp with one hand and puts the idle hand to work with some hamstring stretching. She slings a rolled-up towel around the sole of her foot like a stirrup and grasps the two ends and pulls—a gentle static stretch. "I do every part of my face as well," she says. That, plus plenty of water, she credits with keeping wrinkles at bay.

The program, which takes ninety minutes, started almost by accident a decade ago. All that water was waking her up in the night to pee. Unable to get back to sleep and feeling the time burning away unproductively, she began experimenting with reflexology on her feet. Over the next few weeks, the program expanded north. She now considers her OK technique a key piece, maybe *the* key piece, of the puzzle of her youthfulness.

Is it?

Self-massage has experienced a minirenaissance since distance runners discovered that kneading the soles of their feet seems to relieve painful plantar fasciitis. Ryan Hall, a top U.S. marathon runner, cooked up for himself a "self-therapy" technique that's actually quite similar to Olga's. Instead of a wine bottle, he rolls his back and hips over a hard rubber lacrosse ball. Instead of a towel to provide resistance on those leg lifts, Hall uses a rubber band. And like Olga, he spends a lot of time on his feet.

The fascia—that nerve-rich net of connective tissue—isn't just in our feet. It surrounds every muscle and organ and "participates in every body movement," as the German neurophysiologist Robert Schleip puts it. It basically shrink-wraps us from scalp to toe. Since the fascia also contains a small number of muscle cells, some massage therapists now think the name of the game is to get those muscle cells in the fascia, which have contracted in their tightly wound clients, to let go. There are foam rollers made for that express purpose: they do pretty much the same job as Olga's green bottle.

Schleip thinks massage is the forgotten stepchild of exercise training, lost in the shadow of the big three—cardio, strength, and coordination. That's a weird oversight, since many sports injuries involve overloads of the connective tissue: ligaments, tendons, joint capsules. Massage and gentle stretching are thought to keep those parts in supple performance shape.

If you put the connective tissue of old people under a microscope, it looks different from that of their younger counterparts, Schleip notes. Gone are the wavy undulations (called "crimp") between the collagen fibers that give young connective tissue its resilience. Age runs over connective tissue like a truck backing over cardboard. That's mostly why our gait changes as we age. It's not muscles shortening or tightening: it's the actual fibers getting less bouncy. Proper stretching and deep-tissue massage can restore that old rebound somewhat.

Me, I get a massage every fifteen years whether I feel the need or not. But I've seen deep-tissue massage flat-out save the careers of athletes I've known, and with Olga's forceful endorsement in mind I recently found myself facedown on a foam table in Seattle trying to be open to the experience of a gentleman applying his thumbs to my calves with roughly the pressure of someone standing on them in stiletto heels.

He shook his head as his fingers navigated the bumps. "You know, if you stretched, and got a massage even semi-regularly, you would have a different body." Of course he would say that: massage is his job. But the experience did make me wonder whether massage really can produce those kinds of profound rejuvenating benefits—or at least produce them in people not named Olga Kotelko.

There's little evidence that massage does much for us besides relieve pain. But it clearly has a primal appeal. The firm, predictable, licked-by-momma's-tongue reassurance of massage is something people crave almost from the get-go. Studies have shown that preemie babies who receive massage grow and develop more quickly and stay healthier than babies who don't. Nature made touch feel good so we'd be encouraged to have sex, it is often surmised. But many people find a skillful massage better than sex. They roll off the table, eyes swollen from the release of histamines, almost postcoitally jellylike—really in no shape to drive. (If *bodily* benefits of massage are hard to prove, *mood* benefits aren't. A 2004 meta-analysis found massage reduced depression and anxiety as much as psychotherapy did.)

That kind of deep relaxation may have a ripple effect on the immune system. A recent small study by researchers in the Department of Psychiatry and Behavioral Neuroscience at Cedars-Sinai Medical Center in California found that people who got Swedish massage, similar in some ways to what Olga does, showed increased levels of circulating lymphocytes and decreased levels of the cytokines interleukin and interferon gamma—hallmarks of boosted immune system function.

My friend Juliana Leahy, a massage therapist who once worked on Bob Hope, was excited by the possibilities. Eight hours is a long stretch to be more or less inert, and many older folk roll out of bed feeling cadaverously stiff. To break up the night with a lubricative session of massage and stretching, well, "I think Olga may be on to something," she says.

Olga swears her OK technique is forestalling aging—keeping her looking young and clearing her body's "energy meridians" to let the life force (what the Chinese call chi) surge. She credits it the way Bob Hope credited his daily massage for his breaking 100—in life if not often on the golf course. A sample of two isn't much to go on. But some of Olga's results are hard to ignore.

One afternoon in Montreal last fall, as a spectacular double rainbow lit the sky over the Lachine Canal, Olga began rolling out dough on the counter of Hepple and Taivassalo's apartment. Six dinner guests were coming to learn to make pierogies. Olga relished the idea of imparting a few generations' worth of kitchen wisdom, from a time before pierogies were spat out by machines, perfect and ordinary. It felt like we were learning a dying language.

As the smell of caramelized onions filled the room, Olga took her little squares in hand and zipped them up around the filling with her index finger and the edge of her thumb. Her hands moved like a nightclub magician's. It occurred to me that her face had aged a little in the three years I'd known her, but her hands hadn't. They're young hands.

And then I realized that this isn't normal. Repetitive fine movements, when you do them decade upon decade, tend to bring on joint and tendon pain or both. Arthritis is a real problem for masters athletes—and track and field is among the sports most definitively linked to it. You can see the accommodation to the discomfort in the way competitors hold javelins and hammers. But Olga feels no pain at all—not tonight, and not after the epic digital workouts she gets every other Tuesday, when she helps fold *five hundred dozen* pierogies in her church basement for eight hours straight.

"People ask me why I don't have arthritis," she said that night. "And the thing is, I used to." Every fall, when the temperature dropped and the humidity climbed, her fingertips would go white and numb, she recalled. They'd hurt like the bejeezus when the circulation returned. It's an affliction with a name—Raynaud's disease (or simply "white finger")—and it's caused by spasming blood vessels. Around 5 percent of the population is thought to be affected. There is no known cure.

Except that, in Olga, it went away as soon as she discovered this: From her pocket she produced a couple of firm red sponge balls, like clown noses. In the middle of the night, after carefully massaging her fingers from tip to palm, she will insert a ball between adjacent fingers, and then scissor the fingers closed against the resistance. It's harder than it looks. "The first time you try the balls usually squirt out," she said. And sure enough, when one of the guests, sprinter Colin Field, put down his fork to give it a go, they boinged across the room.

Olga may be out ahead of the science with those little red balls, or else the balls are getting credit for coincidental healing that has nothing to do with them. Either way, chasing them across the room results in some good exercise.

I'll Have What She's Having

People are intensely curious about Olga's diet. It's probably the second most common question she's asked, after the generic "What's your secret?"

That diet must, people assume, be everything the sugar-and-starch-laden standard American diet (aka SAD) isn't. It must feature instead the kinds of foods that are thought to fight disease and forestall aging—all those antioxidants and omega-3 fatty acids and resveratrol.

It must be extraordinarily low in cholesterol. Not just on the evidence of her untroubled heart, but because she's so mentally sharp. (We now know that diets high in cholesterol make us about 50 percent more likely of developing Alzheimer's or dementia.)

She probably avoids dairy and maybe gluten, too, because these promote inflammation, and inflammation contributes to aging. Right?

And maybe she periodically lets herself get really hungry, to kindle stress-response mechanisms and maybe boost her immune system.

Could be she flat-out doesn't eat much. After all, there is only one diet strategy known for certain to extend the human life span: eating 30 or 40 percent less than everyone else. (That gambit, sometimes called CRONyism—for "caloric restriction, optimal nutrition," is cultishly followed by a small group known, for obvious reasons, as "the skinnies.")

Well, here are the facts.

Olga eats a lot, and she eats promiscuously. "I don't believe in diets," she says. I once put the hypothetical to her: If she could extend her life 15 percent by going on a severely calorie-restricted diet, would she do it? Olga answered with an expression that said, "Are you flat-out barking mad?"

She eats what she likes, and apart from sushi there's not much she doesn't like.

She likes her meat and she likes it rare. She eats a lot of protein, because she craves it. I've seen her put back a meat soup, followed by a burger with two additional meat toppings.

Some things we now recognize as superfoods Olga happens to eat, more by preference than design. Bananas, for instance. Olga loves them—not for the nutrition so much as the zippered convenience of them. They're easy snacks before and after competition.

She likes her dairy. Cottage cheese, sour cream, and yogurt are staples. She loves fermented food and always has—a taste developed from her childhood, "because we didn't have refrigeration." Fermenting was a way to preserve food, and so her family would make sauerkraut—a layer of cabbage, a layer of salt—and put it in the garage and leave it till winter, when it would freeze and you'd have to chop out tiles of it to eat. Homemade cottage cheese was easy. While I'd love to report that her family put a frog in every bucket of milk, an old Russian trick to keep it from going sour, they didn't. Because sour was sometimes what you wanted.

"You'd let milk go off—two days at room temperature would do it—then heat it up till it separated and curdled," she says. At any given time on the farm, dill pickles were fermenting in sixty-gallon wooden barrels in the basement—salt, dill, and washed cukes, that's it.

She doesn't skimp on carbs: whole-grain toast and, of course, her beloved Ukrainian pierogies.

She has eaten oatmeal just about every morning for ninety years.

She takes on *a lot* of water. Her belief in hydration verges on religious. I once saw her speaking to a group of kinesiology students. "Forget expensive antiaging creams!" she said, channeling Ron Popeil. "Go get a glass of water right now!" Hold the ice, though. Too much cold makes her digestive system work harder than it needs to, she believes, and brings down her body temperature. "Water, no ice," as a directive to waitstaff, is Olga's version of "Shaken, not stirred"—a defining rule, almost a calling card.

She takes a baby aspirin a day (a simple act that, done religiously, "reduces risk of dying from common cancers by between 10 and 60 percent," a 2009 study published in the *Lancet* suggests) and a multivitamin in winter when the hometown sun can disappear for whole months, but that's where her supplements begin and end. She doesn't believe in them. Anything you can't get in a normal balanced diet, she maintains, you probably don't really need. She trusts Big Nature over Big Pharma, on the principle that whatever youth-promoting supplements we might gobble—from antioxidants to testosterone—the body generally makes better and for free.

If there's one thing that marks Olga's diet it's this: you won't find much processed food in it. It approaches a preindustrial diet, in the sense that there's very little white stuff: not much refined sugar or flour. (It isn't a "paleo" diet because, well, a paleo diet no longer exists. Much of the nutrition our ancestors got from fruits and vegetables has been bred out of even the best-quality supermarket produce, as Jo Robinson, author of *Eating on the Wild Side,* points out.)

She routinely eats "a rainbow," as my kids would say of the colored vegetables we're supposed to have, for the healthful phytochemicals in them. These she grows in her garden: the hundred-foot diet, if you like. Organic to boot.

She eats a lot of the foods we've been eating for centuries, and from these derives benefits we're often just learning of. For example, in the beef and nuts she eats there's zinc. Zinc is good; it's thought to contribute to the prevention of Alzheimer's by maintaining the health of an important neuroprotein called tau.

The fermented food she loves is a source of probiotic bacteria, which aids digestion and is believed by some to boost the immune system. It may even be contributing to her sunny demeanor. Some evidence suggests the probiotic bacteria acts as a kind of natural Prozac.

Those leafy greens from the garden may be contributing to her strength. People have been shown to perform better on strength and endurance tests if they get enough antioxidants and vitamin D.

And by avoiding refined grains and other "high-glycemic" foods that the body quickly converts to sugar, she manages her insulin levels. Processed and refined foods, the bane of the modern diet, jack up insulin levels and ultimately make us fat and sick.

In short, Olga's diet is fairly healthy. But it's not textbook. There are things that get into her that would make purists turn pale.

She eats, for example, immoderate amounts of baked tapioca pudding. I once saw her glop onto her breakfast plate some mysterious "sausage gravy" that looked practically radioactive— because she was just too curious not to. After a few bites she delivered the verdict: "I think I'll live."

This will all, no doubt, be disappointing to people hunting

for specific dietary tips. The truth is, few dimensions of longevity are more bewildering than diet.

Olga's friend Ruth Frith, a shot put specialist from Australia
who competes in field events in the women's *100-plus* category,
eats no vegetables; she can't stand them. Ugo Sansonetti claims
to be fueled mostly by cow's milk ("preferably still warm from
the cow, but cold if necessary"). Californian Bert Morrow, the
world's oldest competitive hurdler until his recent death at age
97, was so committed to bananas that Chiquita signed him to a
three-year ad contract. The world's oldest marathoner, 101-year-old
Fauja Singh, swears by ginger curry. Ushi Okushima, the Oki-
nawan centenarian who's the star of Dan Buettner's book *The
Blue Zones,* can't be without mugwort sake.

The problem with copying the quirky food habits of a really
old stud athlete in order to try to live that long and well is that
everyone is constitutionally different. You or I could drink mugwort sake by the barrel with no salutary effect at all. Our bodies
metabolize and use different foods in unique ways we're only
just beginning to understand. For instance, Olga herself doesn't
benefit all that much from the classically healthy "Mediterranean-
style" diet, high in olive oil and other monounsaturated fats. She
doesn't have the genes for it. (Many people of European descent
carry at least one variant of a gene called PPARgamma that regulates fatty acid storage and glucose metabolism. And these lucky
folk become healthy and svelte as they eat monounsaturated
fats. Olga doesn't have it. And neither, sadly, do I.)

There's another consideration. It could be that Olga is getting some help not just from her own diet but from the diet of
her parents and *their* parents—all that healthful fermented
Ukrainian sauerkraut working its way down generations of
DNA. We know that our diets, like any other environmental
input, can tweak gene expression: that's epigenetics. Scientists
long believed the changes affected only the current organism,

that however your diet changed you, for good or ill, *affected only you*. All epigenetic marks were thought to have been erased in the new embryonic cells of your children.

We now know that's not true. Dietary deficiencies (such as among pregnant mothers in times of famine) can show up as inherited health issues in their children. And conversely new evidence suggests that eating a diet high in healthy ingredients, such as omega-3 fatty acids, may confer advantages on our children and grandchildren. (Early animal studies, such as one published in *Nature* by Stanford scientists in 2011, found beneficial effects from a healthy diet three generations down in a species of worm—prompting one of the lead researchers to speculate on the eventual possibility of a "'super baby' with long life and [a lower risk] of diabetes and metabolic syndrome.") These recent findings up the ante in our own dietary choices.

Not long ago, I started playing around with the "survival curves" in actuarial tables, and discovered some interesting things about what makes people live long lives. There's nothing that makes expected life span fluctuate like diet. Right now I have about a 3 percent chance of living to Olga's age. But when I plugged in different numbers in the food and beverage column, the picture changed dramatically. If I drank a bit more alcohol, and made sure I got my five fruits and veggies a day, I could bump my odds from 3 percent to 17 percent.

What's missing from those curves, though, is our history with food—with bad food. In other words, it's not just what you eat but what you have eaten. We're now learning that childhood diets may have an outsize effect on future health.

My diet growing up was roughly half carbs and half high-fructose corn syrup. While Olga was eating organic tomatoes, I ate a lot of ketchup.

A sort of stockpiling approach to vegetables was my guide—if you forced them down in bulk every couple of days you could coast on their benefits. (Not true, and not unlike my approach to exercise, come to think of it.) A lot of this adds up to massive oxidative stress—a component of premature aging.

It would be a huge stretch to suggest that Olga's diet is the secret of her longevity and her strength. But it is not so much *what* Olga eats as *how* she eats that sets her apart. Her *approach* to food may be an overlooked piece of the puzzle. She eats a good breakfast of steel-cut oatmeal, roughly at the same time each morning. *When* we eat affects our biological clock. A consistent morning feed, Nerina Ramlakhan believes, "reassures your body that in your world there is an adequate supply of food, so it can relax and fall into sleep mode when it needs to."

Olga eats "when I'm hungry," which usually amounts to small meals four to five times a day.

One theory of obesity implicates social and cultural cues in poor food choices. Commercial voices drown out the body's own little voice asking for what it needs. Olga is pretty clear of that kind of pressure. Modern consumer culture isn't pulling her strings very much.

She likes to eat locally sourced food, but not for environmental reasons. What comes out of her garden is cheaper and tastes better. And when she's traveling, she likes to plug into the story of that locale by eating what has always been eaten there.

One of her sweetest memories from the recent world indoor championships in Jyväskylä, Finland, doesn't involve anything that happened on the track. It was a dinner she was treated to by Tanja Taivassalo and her dad at a Finnish restaurant. Below street level they trooped, past bearskin rugs on the walls and great raw wood posts befitting a Viking fort. The servers wore traditional Finnish costumes and laid out dizzying options, beginning with a cinnamon whiskey shot called "dragon breath." ("Are you drinking?" Tanja asked Olga, knowing she was in the middle of her competition. "Of course I am!" Olga replied.) There was smoked herring and reindeer meatballs and lutefisk. And, for dessert, ice cream topped with red currants. But this ice cream was made from tar—the same stuff used on the roads, more or less. For locals the smell of tar is the smell of Mother Finland. Tar is widely used in commercial products such as soaps and shampoos. (The taste must have stirred a vestigial genetic memory in Taivassalo—a Finn—for she pronounced it her favorite dessert ever and would return to the restaurant another day just for another go at it.) To most visitors tar tastes like . . . tar. But Olga found the tar-flavored ice cream close to perfect. Her body approved of it not for its nutrients but for its soul.

"When your body tells you what it wants to eat, does it tune in to the local coordinates?" I asked her one night in Chicago. "Absolutely," she said. "Everything comes into play: the atmosphere, the surroundings, the company." We were having dinner at a local institution called Benny's Chop House. Beneath its veneer of civility—fashionably low light and crisp white tablecloths and waiters as friendly as long-lost relatives—Benny's vibrated with echoes of the stockyards, the trains funneling into them and out, Sinatra singing their praises. Olga knifed into a filet mignon the size of a baseball. She took a bite and closed her eyes.

It's impossible not to think how different Olga's approach to food is from that of Ugo Sansonetti, the Italian sprinter. Sansonetti, who is actually a shade older than Olga, follows a routine that makes the U.S. Marines' position on routine look flighty. Ugo has eaten the same breakfast every morning for eighty-five years: "an egg beaten with sugar, bread with jam." Then half a liter of that still-warm-from-the-cow milk he prefers. Routine is good in many ways—as we'll see in a moment. But to Olga, that kind of robotic consistency in your diet denies the variability of life itself.

In Olga's view, there's a time to eat healthily and a time to let things slide. So, for example, to quaff a wheatgrass shake at a baseball game is wrong on a lot of levels. When she goes to watch the local AA club, the Canadians, play on a sunny afternoon, she will order the proper repast: a hot dog and a beer.

Sometimes food is just something that sustains you on the fly. There are times we have had to bag a sandwich to eat in heavy traffic. That would be anathema in some cultures, but in those moments it works for Olga, who does not like to waste time.

EFFICIENCY is at the heart of another important habit: opportunism.

One day Olga and I were looking over her athletic records when a big discrepancy jumped out. At one meet, she triple-jumped 6.67 meters. Then, just a couple months later, on the other side of the world, she jumped 8.04 meters. "How come I can jump that far on one day and not another day?" she wondered. "How do you explain that?" She hadn't mailed in that first performance: it was the best she could do on that day. But that first meet was a regional competition in her home province. The second was the World Masters Games in Sydney. Again and again we noticed this pattern. The bigger the stage, the more important the moment, the better Olga performs. She is masters track's equivalent of a clutch player.

Adrenaline will do that for you. But Olga wasn't really aware of being more nervous in Sydney, of "bringing more" that day. At some level her body just seems to be telling her: You must use your energy wisely. You cannot just boil it away, as you did in your youth. Pick your battles. Be opportunistic. Some moments in life are loaded and some aren't. People who can tell the difference tend to thrive.

Olga figured this out quite late. As a kid she would stubbornly see every commitment through to completion. She would finish what she started, because this is what a Saskatchewan farm girl did. That ethic changed as she became an adult—and I think the turning point may have been in 1956, on the day she walked out of her marriage. (All Olga ever says about her marriage is that it "failed"—no blame assigned.) Not all tasks are worth bringing your best effort to; not all tasks are worth seeing through to the end. There's no shame in occasionally taking a mulligan.

I was shocked when, as she and I arrived at O'Hare airport on one of our excursions, she requested a wheelchair. Are you kidding? *I* was a better candidate for a wheelchair than she was. "Why not?" she said. Why not conserve her strength for something more productive—and anything is more productive than hauling a suitcase through the rat maze of one of the world's busiest airports.

Routines

Gary Stenlund is the reigning world masters javelin champion among men 70 and up. He turns heads in grocery store lines for the same reason Earl Fee does. It's the incongruity of an older man's head fused to a young man's buff body.

Stenlund, who divides his time between Battle Ground, Washington, and a small coffee farm he runs in Costa Rica, is so dominant at what he does that no one his age has beaten him in competition in almost two decades. At the 2011 outdoor world championships, he launched a winning throw that landed twenty feet farther than his nearest challenger's.

Stenlund heaps credit on a ritual that grounds him and quiets him.

Five mornings a week he finds a tranquil place and performs his "fountain of youth" exercises. It's a program based on an ancient program said to have been hatched by monks in a remote

Tibetan monastery and brought to the West by a retired British army colonel. (The story of this discovery, and the colonel's rejuvenation when he followed it, is told in a New Agey book with a cult following. The newest edition comes with a warning: "This book is not for everyone. You should read it only if you accept the preposterous notion that aging can be reversed.")

The regime—just five movements—marries gymnastics, yoga, and the celebration ritual of a little kid or a lottery winner: spinning in circles with arms outstretched.

The leg lifts and the yoga postures look familiar. The dervish-like spinning is radical. "The spinning is the most important part," Stenlund told me—not because it speeds up the body's "energy vortexes" or any other mumbo jumbo the book offers as explanation, but because it works the vestibular system, perhaps helping keep it in tune. "How do old people get hurt? They lose their balance and fall down," he said.

But to Stenlund, the chief value of the program is as a kind of meditation—just as it was for the monks.

"The monks didn't do it for exercise—they got enough exercise working hard in the fields and maintaining the monastery," he said. "No, they did it to calm themselves down." Stenlund, too. "I'm an animal thrower, an instinct thrower. I do better when I don't get so excited."

The Dalai Lama was once asked the secret of happiness, and he replied without pausing: "Routines." Routines make athletes happy, too, because routines are the foundation of discipline, and discipline delivers results.

Ugo Sansonetti's whole life is a suite of routines almost as idiosyncratic as Stenlund's. If they came with a marketing line, it would be: "You can take the man out of the jungle, but you can't take the jungle out of the man."

When leaving the Costa Rican rain forest in the early 1970s and settling in Rome, Sansonetti brought his bush habits with him. To wash, he sets a little basin inside the bathtub and steps into it, splashing water over himself. ("I use cold water," he says, "to make the body react.") He alternates very cold showers with very hot ones—to keep the body's "thermo-regulatory system" sharp. He sleeps on a hard mattress, simulating his grass jungle bed. He keeps the windows open all winter.

Olga, too, has her routines. Apart from her nighttime OK stretches, there's a ritual she performs on competition days. She lies on her back and puts her legs up against the wall, at a forty-five-degree angle, for thirty minutes. (She's not entirely sure why this works to calm her, only that it does.)

To some degree, it appears not to really matter what our habits are, so long as we have some. Routines are widely considered essential to productivity, not just in sports but in life. They are at the root of positive behavior change, be it weight loss or money management. The trick of routines is that they take the vagaries of motivation off the table.

"What you have to do," says Christa Bortignon, the sprinter, "is just go to the track. It's not really a matter of whether you feel like it or not." Once you get yourself there, your brain entertains the notion that you *must* feel like it because, well, here you are. "Your body knows," Bortignon says, "that when you go into the routine, something's going to happen."

Bortignon slips on headphones and pushes PLAY on Mozart's Piano Concerto No. 21—the stirring *Elvira Madigan* suite, made popular by the suicidally bleak 1967 Swedish film about a tightrope walker. She puts it on a repeating loop. Then she laces up her shoes. "Something different happens when you lace up your shoes," she says.

But as important as routines are in getting things done, too much routine can be stultifying. Indeed, growth comes when we *break* routine, when we confuse our body and brains, jolting them out of the lazily efficient shortcuts they have developed, forcing them to adapt. A good example is walking. By all means develop a habit of walking, says former NASA life-sciences director Joan Vernikos. "But when you walk, don't walk at a steady pace: walk fast and then slower, fast and then slower."

Routine may be a stealth factor in mortality statistics. We know, for instance, that grief kills. While researchers have debated the causes, the physiological dimensions of the broken heart, David Agus, the American physician and personal-genomics pioneer, suspects a main explanation could be something simple. The life of grievers is marked by *disruption of patterns*. Their usual routines often are turned upside down.

Agus sees routine as a key, and largely unheralded, dimension of health and longevity. One of the first things he asks patients who feel sick but have unclear symptoms is this: How's your routine? Any recent disruptions to the time you eat, sleep, exercise, work? For some people these things are in wild flux, and these are the folks who tend, Agus has noticed, to get ill. The human body thrives on routine, and "irregularity," he notes in his recent book *The End of Illness*, "is a big source of physiological stress."

To call Olga a "creature of routine" is not *quite* accurate. In some dimensions of her life she's open to improvisatory flourishes, as the mood strikes. But those riffs are grounded in, and always return to, a steady beat of habits. She has her daily OK exercises and her thrice-weekly aquafit, her Tuesday pierogi-making sessions and her morning oatmeal, her clockwork churchgoing and her nonnegotiable sleep times.

What Olga really has is the optimal mix of routine and flexibility within it. Her reflex, when offered the chance to try something new, is to say "Yes!"—but this impulse then often collides with her self-preservation and self-care instincts, which want to hew to routine. Sometimes she follows the playbook and sometime she calls an audible.

A life that's run with the ship-tight precision of unending routine doesn't leave many cracks for light to get in. In that sense, you could argue that Ugo Sansonetti has gone too far.

His friend Vittorio Colo certainly believed that.

When Ugo was working on a memoir, he approached Vittorio, who had anchored the 4×100 relay team that forever changed masters athletics. Would his old pal and rival be interested in contributing to the writing? Ugo asked. Ugo sent Vittorio some bits of the manuscript in progress.

Vittorio was in every way Ugo's opposite. Where Ugo is robotically systematic, Vittorio is instinctive and spontaneous. He runs on joie de vivre. Vittorio read a few pages, which confirmed what he suspected, then wrote Ugo back.

"Maestro," Vittorio wrote back affectionately, "it would be hypocritical of me to support your project. I live and plan in a way different from what you are preaching."

The very idea of prescriptions, rules for living, and formulas for success struck Vittorio as laughable. "Forget dogmas. Just like those twelfth-century poets waited to write those few verses *until love inspired them,* so do I, if I am training, not follow any agenda, I do not follow a program. I go where my desire takes me regardless whether I'll throw the discus or jump hurdles. Diet? I do not know what it is. I love minestrone with lard. I hate vitamins. I drink a lot of milk, but I do not stay away from a good genuine grappa. It is useless to go on. I live on my own, I leave the rules to others. Even your rules. All of them."

If life is about spontaneity, then plans get in the way of one's natural development, Vittorio said. "People in their forties listen to you, Ugo," Vittorio said. "And those in their sixties, too. But those in their eighties and nineties should be more prudent."

8

Personality

WHEN WE FINISHED the cognitive testing back in Illinois, Olga was given a snapshot of her brain to take home. At dinner that evening with some scientists from the institute, she laid the picture on the table. Something was bothering her.

"There's a hole in my head," she said, pointing to a dark circle in the middle.

"That's a ventricle," said Chaddock. "That's normal: everyone has a hole there."

"Oh, what a relief," Olga said.

Everyone tucked in. When talk of genes and diet and fast-twitch fibers and brain-derived neurotrophic factor had exhausted itself, and before the waiter arrived with a slice of cake big enough to block the wheel of a parked car, silence settled over the table. Attention turned to Olga. Someone asked her, What did *she* think about why she is the way she is?

"Well," she said. "I don't get mad."

"It's a good thing," said Art Kramer.

"I have a little bit of scotch now and then."

"Nice," said Kramer.

"The other thing is maybe my personality. I don't allow people to have a negative effect on me."

Personality is the X factor in a lot of life's endeavors, sports included. At the top levels, pretty much every athlete is a prime physical specimen, and most are loaded with talent, yet superstars are few. The cream have a je ne sais quoi that boils down to character and spirit. There's a kind of rabid competitiveness we might call the eye of the tiger. There are people who you'd never ask to join you for a leisurely jog because you know they wouldn't enjoy it. "It's gotta hurt, otherwise why do it?" is their creed.

Olga has some personality traits you'd expect must help her on the track. There's the tunnel vision, the tolerance for discomfort. She doesn't complain much and she frankly can't stand complainers. If she finds herself sitting with a group of athletes grumbling about their aches and pains, she will get up and leave.

For athletes, fitness isn't an end in itself but a means to an end—a tool in search of a project, sometimes. For three years, after arthritis stopped her tennis career, Christa Bortignon was miserable, as restless as a cat, stalking a new outlet until she found, through Olga, the thing to scratch the itch: track and field. "She couldn't be idle," Olga says. "And me, too, I couldn't be idle."

But hang on. A "performance" personality is one thing. A "longevity personality" is less intuitive.

The link between temperament and life span was long ago suspected—Darwin was pretty sure that healthy, *happy* people are life's survivors—but only quite recently confirmed. Mentally healthy people live longer. "Positive" people live longer. These facts have become a foundational stone for a whole discipline called "happiness studies." Howard Friedman, a University of California research psychologist and steward of one of the largest and most scientific longitudinal study of Americans, has even isolated in many centenarians what he calls the "self-healing personality"—as distinct from a "disease-prone personality."

The simplest explanation for how a personality trait can make someone live longer is this: the trait drives healthy behaviors. Conscientious people are more likely to wear seat belts. Outgoing people knit a safety net of stable friends and neighbors around them.

Not long ago, Olga sat down to take a personality test called the NEO Five-Factor Inventory-3.

The five-factor model is perhaps the gold standard of personality snapshots. Shorter and simpler than other personality tests, it nonetheless holds up across cultures, and even crosses the species barrier: anthropologists have used it to categorize chimpanzees. Its premise is that, within the limitless subtle hues that make up every human personality, there are five dimensions—five "traits"—that matter most. Taken together they reveal who we essentially *are*. Those traits are openness, conscientiousness, extroversion, agreeableness, and neuroticism. You can remember them by the acronym OCEAN. (Or if you're Canadian, like Olga, CANOE.)

Scientists are interested in how exceptionally long-lived and well-lived people score on the various dimensions of the five-factor test. Repeating patterns are meaningful in somewhat the same way that common patterns of genes in centenarians are meaningful; they hint that there's a recipe book. The difference is that, unlike genes, our personalities we can do something about. It's thought that around 50 percent of our personality is heritable. The rest is shaped by our experience—which is to say, it's open for revision.

OPENNESS

There's a phrase Olga uses so often I've often thought to stencil it on a T-shirt for her: "What have I got to lose?"

People with high O scores, who some data suggest live longer, are curious and broad-minded, wallow happily in abstractions and what-ifs, are piqued rather than threatened by the new, and may be less bound to tradition and deferent to authority.

Openness is a trait more commonly linked to liberals than conservatives, and indeed Olga's responses on the NEO inventory led testers to label her "progressive." (So, *is* political affiliation a predictor of long life? Inconclusive. Liberals, armed with those high O scores, also eat healthier diets. But conservatives have more money, which is linked to better health. And conservatives say they are happier. Overall, it's probably a wash.) Progressives do tend to tolerate ambiguity well—and that trait does correlate with long life, irrespective of your politics.

One of the things Olga is most proud of is how much of the world she has been able to see, despite being too poor for much of her life to afford fancy trips. She saved, found her opportunities, and pulled the trigger. She once did a teacher exchange to Uganda, then tacked on a trip to Bangkok (arranged by a priest who was part of the teaching team and had taken an academic interest in the red-light district there). From there she decided to push forward instead of backtracking—Cairo and Lisbon and Southampton and home—circumnavigating the globe for an additional surcharge of thirty-seven cents.

"What's on your life list?" I asked Olga not long ago. Nothing jumped immediately to her mind. "I'm satisfied with what I've done," she said. It reminded me what the gerontologist Karl Pillemer turned up in his systematic collection of practical advice from older Americans. There is no one "secret" of happiness, Pillemer found, but there is a secret of unhappiness, and that is regret. Olga has no real regrets, because she went for it. Her bucket list is blank. Except for one thing.

"I'm going to learn to play the piano," she said. There's an upright piano in the house sitting idle. Olga got it for her daughters, but for reasons she blames herself for (she didn't find the secret of motivating them without pushing them) they didn't stick with it. Maybe she can find a good teacher in the neighborhood, Olga figures. Or teach herself.

OLGA'S OPENNESS SCORE: HIGH.

(NEO test scores follow the normal bell curve. Around a quarter of us will score "high" on a given dimension. Around 7 percent will score "very high.")

CONSCIENTIOUSNESS

Conscientiousness is really a measure of goal orientedness. It's about the steady march toward your set objectives—hardships be damned—and often the high moral road taken to get there. Conscientious people, as Costa and McRae, inventors of the five-factor model, frame it, "are able to tolerate unsatisfied needs without abandoning their plan of action."

High C-score people? You borrowed their notes when you missed class. They had an investment portfolio when you were still impulsively buying comic books. They dust the tops of doors.

Friedman and Martin, directors of the Longevity Project, found conscientiousness to be the one trait that unambiguously predicted long life. It may be the most complex of the five, and in many of its facets you can see Olga reflected. "Detail oriented"? When historians one day want to survey her records, the primary source won't be any governing body of sport. It'll be Olga's own logbooks, with scores in every event she has competed in, going back a decade, written in longhand.

And then there's her habit of self-correction, which makes Benjamin Franklin's efforts in this respect seem halfhearted. *Fail. Appraise the failure. Fix.* There's a line she likes—"a stumble may prevent a fall." She understands it to mean that a little error may preclude a later, disastrous one. So welcome the little error. And then learn from it. Scientists who study high performance have found Olga's strategy to be sound: we get better faster when we have immediate feedback, so that we can tweak what we did wrong and positively reinforce what we did right—and that applies whether a coach holds up the mirror to us or we conscientiously find a way to hold it up to ourselves. The margins of her puzzle books are peppered with private notes, such as "Careless errors!" and "Getting better—or are they getting easier?" plus the occasional star for cracking a tough nut. In her bowling league, after each strike, she takes note of where she was standing and how hard she threw "and then I try to duplicate those conditions."

One of the highest compliments Olga gives fellow masters athletes is that they are "honest" competitors. Carol LaFayette: "She's an honest runner," says Olga. I pressed her on what that meant. "It means she's serious about her athletics, she's committed, she puts in the time."

The social dimension of conscientiousness often gets forgotten. Conscientious people care about keeping up their social networks, and there are good practical reasons to do so. Friends are a buffer against unexpected hardship; they're money in the bank. Olga has at least three banks: church, family, and the track community. A solid reputation in such circles is gold-backed currency.

When conscientious people reach the goals they've so methodically homed in on, they are happy. And happy people tend to go on living.

OLGA'S CONSCIENTIOUSNESS SCORE: HIGH.

EXTROVERSION

An offshoot of the New England Centenarian Study looked one generation ahead, at the children of centenarians. What personality traits might the kids have inherited from their Super Senior parents?

Turns out a lot of them were extroverts.

(What is an extrovert? Here's a good rule of thumb. If you walk away from a party with more energy than when you went in, you're an extrovert. If you walk away with less, you're an introvert.)

Throughout history, extroverts got an evolutionary leg up. They were the ones with the energy and will to forge helpful alliances and make useful discoveries, out there in a world full of scary other people.

There's evidence that extroverts live longer, and theories abound about why that is, from lower levels of stress chemicals to higher levels of physical activity. High extroversion levels, notes Thomas Perls, head of the New England Centenarian Study, are also associated with "looking after yourself."

I'm not convinced Olga is a true extrovert. I think she's a combination, as most of us are. There are times when she's overpeopled and needs to retreat into solitude. (In some ways her living arrangements, in the same house as her daughter and son-in-law but separate from them, are perfect. It's a bit like Thoreau's setup in the cabin at Walden Pond. If things got *too* quiet he knew he could always go to his mom's for a hot dinner.)

But plenty of times I have seen that impulse to reach out bloom in Olga.

At the javelin event at the world indoor championships in Kamloops, when her five throws were done and all the other competitors had left the pitch, Olga sought out the officials to thank them, arms outstretched. At one point she buttonholed a guy taking down a camera tripod. "I wasn't officiating!" he protested, then gave a what-the-hell gesture and hugged Olga anyway.

In recent years she has developed a ritual that is two parts generosity and one part housecleaning. To people she meets whom she has hit it off with, or has been impressed by, she gives one of her medals. There is a funny Facebook posting from the captain of the Holland America cruise liner *Eurodam,* registering his astonishment after a 90-year-old woman, who stood only as high as his breast pocket, pressed into his hand the gold medal she won by running a sub-24-second 100-meter in Finland: his to keep.

On balance, the NEO testers concluded that Olga is "likely to thrive with people around."

OLGA'S EXTROVERSION SCORE: HIGH.

AGREEABLENESS

People with high A scores have lots of friends and very few enemies. They listen well and are generous with praise and affection. (A good test of character, it is often said, is to watch how people treat someone who can do them no obvious good. Such people would score high in agreeableness.) There's a simple openheartedness to Olga that those who have met her appreciate. Such a trait is hard to fake.

In the German documentary film *Autumn Gold,* we meet 94-year-old Italian athlete Ljubica "Gabre" Gabric as she trains for the 2009 world outdoor championships in Lahti, Finland.

Gabric is depressed. She competed for Italy in the 1936 Summer Olympics. Her body remembers the best efforts of those earlier years, and her best efforts now are so far off they're driving her crazy. She can barely throw and run and jump a third as well as she once could.

What's worse, she thought she had the 90-and-over division to herself. But now she is hearing word of a new entrant who's breaking records left and right and leaving spike prints on the carcasses of the vanquished.

Cut to: Olga, smiling sweetly.

Gabric finishes second to Olga in the shot put, and later the camera finds her standing alone, dejected. Olga enters the frame. She puts a hand gently on Gabric's shoulder. "Gabby, you did great!" Gabric, you can tell, is unsure how to respond. The comment could be read as the ultimate backhanded compliment, a bit of veiled arrogance, since Olga obviously did even "greater."

Gabric points out that Olga smoked her.

"But, Gabby, I'm younger!" Olga says. "I'm only ninety, and you're ninety-four." Gabric's face hardens. "She was pissed off," the filmmaker Jan Tenhaven recalls. Gabric had taken great pains, throughout the entire shooting of the film, not to reveal her exact age; she was the only competitor to keep it a secret. "She was so afraid of people thinking she was old," Tenhaven says. "If people knew she was ninety-four they might try to, you know, help her across the street." Now she has been busted, live on camera.

And then, on the brink of social disaster, Olga performs a Houdini-like escape. Gabric is, in fact, the best 94-year-old in the world, Olga notes, and she shouldn't forget it. And she is so plainly sincere that Gabric's body language changes. Her defenses start to melt. A hint of a smile crosses her face.

The Yiddish word "mensch" literally just means "human being," but figuratively it implies someone for whom the best traits of our species—kindness and generosity—emerge reflexively, every day.

Of all the NEO Five-Factor traits, agreeableness is the one Olga scores highest in. Indeed, she is so promiscuous with the phrase "I love you" that for a long time I found it irksome. It seems to freeze those who receive it after they have more or less just met her. (Should they reciprocate? Is it rude not to?) But I came to believe it's an authentic and actually quite touching habit, not so different from a new friend telling you in a word that they "honor the God" in you. *Namaste.*

Not all old folk are mensches, to be sure. Indeed, some psychologists have suggested that people's personalities can become more astringent as they age. The weaker we get, the more vulnerable we feel, the crustier we have to be to survive, goes the theory. Writer Chris Crowley evoked the image of a "mangy old wolf, snarling furiously at the slightest threat." But the spirit of menschness is alive in the world of masters track—especially in the upper registers—even among the ferociously competitive top seeds.

In the early 1990s, psychologists at the University of Akron studied a group of seniors, aged 66 to 101, who had outlived their siblings by seven years on average, and found one agreeable trait in particular emerged as the difference: sense of humor. The surviving sib reportedly had a better one.

Let's clarify: are we talking life-of-the-party funny or "good-humored"? Olga is definitely not the first, but she is the second. Funny-ha-ha people can be prickly. Comedians often aren't all that "nice," and as a group they tend not to be particularly long-lived, as Hannah Holmes points out in her fine book *Quirk*. Funny people court conflict (the comic Louis CK has reduced his sensitivity to offending people to the point where it's "just gone, like an organ that has been removed"), while agreeable people skirt it—again an actuarially beneficial thing. One small study of Greek centenarians found that pretty much all of them avoided people who stressed them out or brought them down. When conflict boiled up, they basically fled.

This is Olga.

Nice is also good in a practical sense. Nice works. People want to cooperate with you when you're nice. Psychiatrist George Vaillant, director of the Grant study of the lives of Harvard men, found kindness and generosity actually correlate with financial success.

OLGA'S AGREEABLENESS SCORE: VERY HIGH.

NEUROTICISM

That T-shirt I want to make for Olga that says "What have I got to lose?"—on the back it will say, "If it's meant to be it'll happen." That's the phrase she says second most. It suggests a level of easy fatalism that must be the opposite of neuroticism. Neurotics make life complicated for themselves. They worry about asteroid defense, second-guess their restaurant orders, maybe don't own an answering machine to avoid knowing definitively that no one has called. Neuroticism is the only trait of the five on which you want to score low.

You can think of neuroticism as a snapshot of the way the amygdala is wired. How quickly does it mount a fight-or-flight response to novelty? (Or maybe, more accurately, it's a glimpse of how the amygdala talks to the frontal cortex, and how they sort out between them how to feel about novelty.) What the neurotic sees as a threat, the non-neurotic sees as an opportunity to show the world their best stuff. (The biblical James had a word for that: *endurance*.) The opposite of a neurotic wreck is someone, whether a beach bum or a test pilot, who puts the brakes on stress so quickly that the world never sees him sweat. We call that trait easygoingness, and easygoingness is strongly linked to well-being.

Neurotics make more evolutionary sense than laid-back people—they're sensitive to potential trouble, and may take fewer chances—so it doesn't quite square that they should live shorter lives, and the evidence here is actually mixed. The Longevity Project found that the carefree and unambitious had an *increased* mortality risk, though the New England Centenarian Study found the opposite. In the latter case, not only did those very old subjects themselves tend to be easygoing, so did *their children*. They had low N scores. They knew how to shed stress.

Over his years in the Kenyan bush studying baboons, comparing the behaviors he observed with the endocrine profiles gleaned from blood samples, the Stanford neuroscientist Robert Sapolsky came to realize that the temperament of the baboon mattered a lot to its health. "You could be the highest-ranking guy on the block," Sapolsky noted recently, "but if you don't have these particular personality styles you're going to have just as crummy a physiology as the number twenty baboon."

The healthiest baboons were, to stretch the anthropomorphizing, the least neurotic ones. They reacted to genuine threats but didn't waste energy imagining provocations that weren't there. They seemed not to fret about issues outside their control. When they did get stung by defeat, they turned to others in the troop for solace and support—bleeding off stress and maintaining strong social bonds in the bargain.

Those unbothered baboons, Sapolsky found, had half the levels of circulating glucocorticoids—stress hormones—as their more neurotic brethren, and lived two to three years longer. That's 10 percent more life, just for being born cool.

OLGA'S NEUROTICISM SCORE: LOW.

Of course, there are some traits relevant to longevity that slip through the NEO net.

What emerged from George Vaillant's study of Harvard men was that "the capacity for intimate relationships" was an important predictor of whether these guys flourished or sank. Note the language: the *capacity* for intimate relationships. As if it's a gift (if innate) or a skill (if learned). If it's learned, the learning is clearly harder for some "types" than for others. Those men who couldn't reach out intimately, from their island, tended to die early.

The primo recipe for surviving and thriving, from Vaillant's data, was the combination of openheartedness with old-school traits such as persistence and orderliness. You're in good shape if you can be "affectionate about people and organized about things," as columnist David Brooks nicely summarized the data. Yin and yang are in balance, and the wheel just keeps on turning.

The most contentious personality trait in all of the longevity studies must be optimism. Some researchers say, with gadfly élan, that it's actually a hindrance to long life. "Cheerful and optimistic children were less likely to live to old age than their more staid and sober counterparts," Howard Friedman found from his subject pool of long livers. But some researchers have found the opposite. Among them is Yale psychologist Becca Levy, who surveyed 660 people in Oxford, Ohio, in 1975. Levy's team contacted the same subjects again in 2002. More of the optimists, whose self-description indicated a fiercer will to live, were still alive.

What does seem patently true is that if you are an optimist, Pollyannaish optimism is probably *not* the best variety to have if you hope to be around to muss your great-grandkids' hair. Recent research points to the advantages of a more nuanced outlook. One study out of the University of California at San Francisco suggested being able to hold conflicting opinions—say, you see the bad but choose to wrap it in the good—promotes health. If true, that bodes well for Olga, because I think she's less a Pollyanna than a realistic optimist. A Pollyanna says: "People are good." A realistic optimist says: "I don't let people have a negative effect on me." Olga spends her thoughts and energy on things that can help her. Her ratio of buoyant positive thoughts to draggy negative ones is about the same as my 9-year-old daughter's: pretty high. She has surveyed the good and bad, and on balance she's happy to be here.

In the airport in Champaign, Illinois, after our visit to the Beckman Institute, she and I moved toward security, where a high-tech full-body scanner blocked our path.

I took off my shoes. She didn't. A sign said that you didn't have to if you're over 75. She started toward the machine.

"Excuse me, ma'am," a security agent said. "How old are you?"

"Ninety-three," Olga replied.

The agent looked at her. It was clear she didn't mean to say this next thing out loud: "You're shitting me." She quickly gathered herself. "Excuse me, ma'am. I'm sorry."

The agent looked away, shook it off, and then looked at Olga again. "You're how old?"

"Ninety-three."

"What's your secret?" she asked finally.

"Enjoy life!" Olga replied.

The agent nodded. A grin infiltrated her face. Then she turned to her supervisor, somewhere behind the barrier. "I quit!" she announced.

THERE is, finally, a trait related to optimism but different enough that it's not really covered in the NEO or any other personality test. That trait is self-belief.

On her first throw of her first event of the world outdoor championships in 2012, Olga threw the hammer 16.71 meters. The mark bettered the previous world record by a full ten feet.

Afterward, she appeared to be a few inches taller. She had no explanation, but then she has stopped needing explanations for the things she does. It was shaping up to be an awesome meet.

Olga's Bob Beamon–like feat was later declared to be a measuring error. The adjusted mark was actually a little shorter than she'd thrown in Sydney. But nobody who learned of the error had the heart to tell her. It might crack her confidence. Confidence—which we might define as a sense that the odds are tipped in our favor—is a delicate thing. Confidence makes stuff happen, as the German psychologist Lysann Damisch recently demonstrated.

She and two colleagues at the University of Cologne designed
a study to investigate the relationship between superstition and
performance. In one trial, when her subjects carried their
"lucky charms" in their pockets, they did better on memory
and dexterity tasks; when testers took their lucky charms away,
they did worse. (The results were published, in 2010, in *Psychological Science*.) The difference, Damisch discovered through a
separate test, was confidence. Magical thinking improves confidence, and confidence improves performance.

The mind elevates the body to heights the body has no
business reaching: that is a truism of sports psychology. "The
brain," Timothy Noakes, a professor of exercise science at the
University of Cape Town and author of *The Lore of Running*,
once said, "is the ultimate determinant of performance." When
we excel, it's because at some level we believed we would, Noakes
thinks. And when we come up short, it's because at some level
we doubted. "There is," Noakes wrote to me in an e-mail, "a
massive placebo effect in everything we do."

A number of ingenious studies have proved this true.

In one, exercise scientists at Northumbria University in
England put athlete-subjects on stationary bikes. In front of
them was a video display of another cyclist, a competitor they
were asked to chase. The competitor was, in fact, themselves—an
"avatar" riding at the top speed the subject was capable of based
on their best time. In the trial, the subjects were able to stay on
the avatar's tail.

Then the testers cheated. They sped up the avatar. Now it
was moving faster than the subjects ever had. Still the subjects
kept up. The testers had coaxed higher performance from the
riders by manipulating the riders' expectations of what they
thought they should be able to do. "It comes back to the belief
system within the athlete," Kevin Thompson, the lead scientist
in the study, told the *New York Times*.

One study by researchers at the Institute of Cognitive Science in France suggests that "affirmative language" helps performance. The brain learns increasingly to respond to our self-coaching. A kind of "volition switch," as the science writer and Ironman veteran Christopher Bergland calls it, in the frontal cortex is activated. "Go!," you bark, and billions of synapses respond to the command. In that moment you can talk your brain into just about anything.

This principle—that beliefs *about ourselves* profoundly affect what we can do—extends vividly beyond the arena of sport.

Psychologist Becca Levy, who studies the effects of ageism, looked at a group of seniors aged 70 to 96. Some had drunk the cultural Kool-Aid (probably from watching too much network TV) that seniors are weak and infirm, and that little should be expected of them. Others had a much more optimistic view of what seniors can do. Levy found actual physiological differences between these two groups. The ones who didn't feel stigmatized about their age showed *less hearing loss* than the others over the thirty-six months that she observed them.

When people think of themselves as capable of continual improvement, they improve. Middle school girls who learned that the brain is plastic, that it grows and adapts—meaning there's no reason people can't get smarter and smarter— suddenly started to achieve higher grades. Stanford developmental psychologist Carol Dweck, who devised that study, calls this the "growth mindset." It is a way of looking at the world, and at ourselves. It is a bit like that "remove grid" function on some desktop publishing programs. With a push of that key, the lines and boundaries that defined your work space vanish. The growth mind-set is about first learning that that key exists, then thinking to push it, and then pushing it. Life becomes limitless.

Is Olga the way she is, at least partly, *because she has decided to be?*

I have repeatedly asked her if she thinks she'll still be competing at 100. There's never any hesitation. She's convinced of it. And we're not talking about merely participating, doddering out there and accepting the laurels they give to the last one standing. She'll be *performing,* and performing well. As she imagines it, so she creates it.

It's no longer even controversial to say that our expectations are powerful medicine. Think of some of Olga's practices: her deep belief in reflexology; her conviction that her stretching routine is clearing energy meridians and promoting vitality; her avoidance of cold things, from iced drinks to drafty rooms, because they lower her body temperature and compromise her immune system. Are these things sound science? In a sense it doesn't matter. What matters is that she believes in them.

Coaches throw around the term "placebo effect" loosely, when what they really mean is something like the power of positive thinking to make us run faster or compete harder or win more often. The placebo effect is best seen as measurable change at the biochemical level. The mere act of swallowing a pill you think will work sets in motion a cascade of processes in the body. Placebos have been found to relieve pain (by triggering natural opioids), modulate heart rate, and suppress inflammation. To believe in yourself deeply is to swallow a sugar pill your mind has concocted. "If you think your body is in top shape, your immune system will more effectively fight back when a bug attacks," the University of Georgia gerontologist Leonard Poon, a leader in the new field of "psychoneuroimmunology," put it not long ago.

The Harvard psychologist Ellen Langer has spent close to four decades studying how people can change their physical health by changing their minds, but she is still best known for an experiment from early in her career, the results of which almost beggar belief.

Langer rounded up a group of test subjects in their seventies and early eighties and invited them to a retreat in rural New Hampshire. They'd spend a week in an old monastery that had been tricked out to create a kind of time warp. The year was 1979, but inside the building every cue—from the decor to the appliances to the newspapers and magazines lying around— suggested a date twenty years earlier. Perry Como crooned on the radio and Ed Sullivan swayed on a black-and-white TV. The books on the shelves were the bestsellers of 1959. Langer's test subjects were shielded from all reminders of their true age. There were no mirrors and the only self-portraits allowed were photos of their much younger selves. They had had to haul their own suitcases up the stairs—just as would have been expected of middle-aged men. Every evening they had lively discussions about "current events" such as the launch of the first U.S. satellite and the merits of the hot new movies, such as *Ben-Hur*, that they'd just watched together. They were counseled not to treat the experience as a lark but to try to buy in. "The sooner you let yourselves go," Langer told them, "the more fun you'll have."

What followed resembled something out of the movie *Cocoon*. After a week, the men were visibly rejuvenated. Objective observers said every one of them looked much younger. Their posture was better. Their hearing and memory and grip strength had improved. They did better on cognitive tests than they had done a week earlier. Their joints had become more flexible; their arthritis pain diminished. Somehow, in the crucible of a powerful thought experiment, stimulating environment, a high level of control over what they were allowed to do, and investigators hanging respectfully on their every word for a whole week, the men had, at least temporarily, turned back the clock.

We obviously can't will ourselves to stop aging entirely. But can we will ourselves to productively use all of our naturally allotted life span, simply refusing the infirmity stage?

Ralph Maxwell, like Olga, has decided he will keep on trucking at a high level. "I am going to race until I am one hundred
and one years old," he said not long ago. "I hope to establish
some world records when I turn ninety-five." I once watched him
complete a decathlon by running the mile—his least favorite
event—in blazing midday California sunshine. To keep himself
going, he sang this ditty to himself, in an endless repeating loop:

I'm one tough SOB
Ain't no one tougher than me
When it gets too tough for the rest, well
It's just about right for me.

It was so hot I grew woozy and had to step into the shade.
Maxwell chugged on around the track, never stopping, to a
personal-best time. At one point I thought I could hear him
humming.

If it's true that, in the case of Olga and Ralph Maxwell and
other Super Seniors, intention to thrive is somehow, partly,
making it so, there may be a rational explanation.

"What it probably boils down to," says Daniel Krupp, a
research psychologist who's also a postdoctoral fellow in math
and statistics at Queen's University, "is that healthier people are
more likely to live longer and are also more likely to *feel like* living longer, because life is inherently more enjoyable when you
feel well. Hence, we find a correlation between the desire to live
and actual life span."

If there remain doubters about the role of will in late-stage
life, the story of Alfred Proksch will test them.

Proksch is one of the five masters athletes featured in Jan
Tenhaven's documentary film *Autumn Gold*. Tenhaven had been
trolling for candidates at the 2007 World Masters Athletics
Championships in Riccione, Italy, when he asked Proksch if

he'd be interested in participating. The Austrian discus thrower weighed the request. "Yes, young man, you can film me," he finally said. Then added, "But I cannot promise I will still be alive at the end of it."

Proksch was then 98 years old. The next outdoor world championships, in Lahti, Finland, toward which the story was building, were still two years away.

But Proksch was in good shape. It became clear that his caveat to Tenhaven was mostly dry Czech humor. He was gunning for Lahti. And all was proceeding apace until one day, approaching his 100th birthday, Proksch fell and broke his knee.

He decided to get a new one. Austria's public-health system defers, in such cases, to the doctor's judgment. "The doctor said he had the heart of a sixty-year-old, so it was no problem," Tenhaven recalls. The filmmaker asked the surgeon if Proksch was the oldest person ever to receive an artificial knee. He hadn't heard of anyone older, the surgeon replied.

The procedure was successful. But a few months later, Proksch suffered a heart attack in his apartment in Vienna, where he lived alone. A neighbor discovered him, clinically dead, and paramedics reanimated him.

This was *three weeks* before the big meet.

So it surprised everyone but himself when Alfred Proksch stepped gingerly onto the infield in Lahti. He was entered in the M100 division. He parked his walker and stepped into the throwing circle. His goal, he said, was to improve on each toss.

"He was so disciplined and he was fighting so hard to make it," Tenhaven recalls. "When he did his third and last throw, he almost fell into a chair, he was so relieved, and you could see the tension drain from his face, and he said, 'This is going to be a brilliant autumn.' Everybody in the crew could feel at this moment that it was the last big thing he did, and now he was ready to die."

9

What Makes Olga Run?

IT'S NOT JUST the body that ages. On bad days the body feels like a proxy for the spirit, the soul, the dreams, those plans to clean the attic. Everything. The body is just the part of you where the years show.

That's why the intrigue surrounding Olga isn't just her records; it's her desire to keep chalking them up. What makes someone continue to be a go-getter at an age when there really isn't a whole lot more to go get? Whatever the constituents of competitive fire are—maybe the oxygen of ego, the accelerant of hormones, and a chip on your shoulder to burn—it's crazily rare to still have them going in your 90s. Generally, everything hurts, money's tight, and the friends and family you used to do it for, and with, are dying off around you. This is the stage where many people raise the white flag, upgrade their TV cable package, and settle in for the sleepy denouement.

Yet here is Olga hours before a meet far from home, still getting butterflies, still feeling the intensity of her task, still bringing the emotion. A few years ago, at the famous Hayward Classic track meet in Eugene, Oregon, Olga got off a javelin toss that felt heroic. But officials disallowed it because she hadn't announced that she was about to throw. "I was so mad," she recalls. "And the next throw was five meters farther."

It's hard enough for many athletes to stay motivated at the end of a long career, race in and race out, when they're being pushed. But when you're far and away the class of the field, you're left to cook up strategies to push yourself.

At the world indoor championships, Nolan Shaheed, the great 60-something American middle-distance runner, was coming off a foot injury but still blew away the field in the 800-meter, just missing the world record. After crossing the line, Shaheed tarried in the finish area, and then he singled out an exhausted laggard who was still down the home stretch. Shaheed walked up to the surprised man and shook his head and thanked him. Shaheed likes to pretend the people he's about to lap are actually ahead of him in the race. This gentleman was that guy for Shaheed, and he helped him find an extra gear.

Olga has a different trick. During a race, she thinks about what it's going to feel like to cross the finish line, so every stride brings her closer to that imagined surge of accomplishment. Relief is at hand, just twenty meters away now, a handful of breaths, one last blast and she's home.

Such microstrategies work for her. But Olga also draws inspiration from the big picture. It's not lost on her that she is lucky to be competing at an incredibly ripe *historical* moment. Even fifty years ago, a woman trying to enter the Boston Marathon would have had her form returned with the explanation that the distance was too strenuous for women. Older joggers were likely to have been chased down by people wondering which nursing home they were escaping from. Just in her adult lifetime, everything has changed. Olga can run. Here, now, she can run. She doesn't have to do it dressed as a man or sweating under a hijab. She can take high-jump practice with 15-year-olds and nobody looks askance. "This is a golden age of Old Age," says Ken Stone of masterstrack.com. "Now older athletes are celebrated, honored, and admired, and Olga's leveraging that. She's *gorging* on the sport, she's gorging on her ability."

Not long ago gerontologists uncovered a quirky fact: scientists who win the Nobel Prize live longer than scientists who don't. The prizes have been given away for long enough that the finding is statistically significant. "Correcting for potential biases," write the authors, "we estimate that winning the Prize, compared to merely being nominated, is associated with between one and two years of extra longevity."

The difference appears to be the very public boost in status. Status, this study suggests, is a life extender. Winning track meets isn't quite like winning a Nobel Prize, but clearly how we regard ourselves is a factor in longevity. Being judged peerless, and having the media duly record that fact, must feel to Olga a little like winning the Nobel, albeit minus the $1.2 million. Feeling like a winner in your 90s is rare and wonderful in a culture that tends to draw the curtains on people over 70.

Winning changes us. Rejuvenates us, even. Social scientists call it "the Winning Effect." It's really more chemical than psychological. Winning boosts testosterone levels, right after the event, as much as losing depresses them. It is a kind of natural hormone-replacement therapy.

For Olga, the past decade has been one big, snowballing win.

In Dorado, Puerto Rico, at age 85, Olga deployed the Western Roll to set a high-jump mark that still stands.

By age 90 she held two dozen world records, but was still cruising somewhat under the radar. At the 2009 World Masters Games in Sydney, "the largest multisport event" the world had ever seen, the face of the games wasn't Olga; it was her much-decorated pal Ruth Frith, who was competing in the field events at age 100. But Frith's W90 field records were now sitting ducks, and by the end of that fortnight Olga had methodically taken them down. One observer called Olga "a machine."

A few months later, in punishing heat at the world championships in Lahti, Finland, Olga bagged five more world records. Reporters from Britain, France, and Brazil were dispatched to Canada to write about her.

At the world indoor championships in Jyväskylä—a university town set among sky-mirroring lakes in central Finland—Olga and Christa Bortignon accounted for almost half of Canada's gold medals.

An extended winning streak carries with it a certain juju. It makes Olga feel young and unvanquishable—which sends her back out there onto the pitch with through-the-roof confidence, with predictably positive results. And so the cycle continues, a feedback loop of age-defeating vitality.

Remember that difference between Olga's performance at small meets and big ones? It's on those biggest stages, when the most eyes are on her, that she performs best. Attention is its own kind of currency, and the winner stands there feigning modesty as it's tossed like rice by an approving crowd. Olga may appear to be the sweet and humble churchgoing lady, "but don't think that she doesn't secretly love being the object of attention and glory," notes Ken Stone.

Remember also the study of British government workers—which found that managers lived longer than their underlings, and invited the explanation that a sense of control is the difference (it's less stressful to boss than to be bossed)? There's a competing explanation. Studies have found a link between professional achievement and self-esteem strong enough "to account for the health gradient. Higher self-esteem, it appears, means lower levels of coronary heart disease risk." Self-esteem levels may even affect immune-system function, some research suggests. And nothing injects self-esteem more reliably than winning, over and over.

Why does Olga enter eleven events when most people enter one or two? The obvious answer is because she can. But there's another angle on this: she is spreading her risk. And spreading risk is a good strategy for emotional health.

Sports psychologists who work with Olympians often find themselves having to delicately manage damaged clients who have been whipsawed by the sudden journey from hero to cipher once their moment of glory had passed. It's even worse if after all that work the athlete never made it to the podium. They went all in. And now it's over, and they are a failure—at least in the terms that they have used to define their life.

Olga does the opposite. She really can't lose. Her multisport program is a kind of hedge fund. She's bound to win at least a few gold medals—even if some hammer throw or 200-meter specialist emerges to eclipse her in those events. So she can always consider herself a winner. (Olga never thinks about that 100-meter record that somebody else holds—unless I annoyingly bring it up.)

The other benefit of going wide is not so much psychological as physiological.

Our resilience depends, you could say, on our weakest piece. For many athletes that's the knees. It could be a hip. Whatever action your sport requires, doing it a million times will pretty much expose the most fragile connector. Olga's MO is "Develop everything." Don't neglect any one part of yourself. There is no one exercise Olga does perfectly or with bottom-of-the-ninth abandon. And maybe that's key. Going wide helps keep her from getting injured.

Masters runners have another trick to stay motivated. They find a way to feel like they are still improving. The key is not to measure yourself against a fixed mark, or against your former self. (This is part of why so few Olympians carry on into seniors. It's not just that their bodies are beaten up; it's that it's too difficult to forget what they used to be capable of.) The key is to measure yourself against realistic expectations for who you are right now.

Ed Whitlock has run a marathon in two hours thirty minutes, and a mile in four minutes thirty seconds. That marathon personal best, recorded when he was 26, put him in the top 1 percent of marathoners—a very, very good accomplishment. But "age-graded" Ed is actually a far better marathoner today. His 3:15 at age 80 puts him on the summit—not just the best current marathoner (relative to age) but *the best marathoner who ever lived*. It's the latter score and not the former that Whitlock identifies with and uses as fuel to continue.

Masters athletics, the way it's set up, almost engineers a positive outlook. Instead of starting to dread birthdays round about age 35, as many of us do, masters athletes quite openly look forward to getting a year older. Because now you're that much closer to moving up a category, whereupon, if you (touch wood) remain healthy, you will get a chance to whip a whole new cohort. Every five years you are reborn.

At a meet I heard Philippa Raschker, the great American masters decathlete, consoling herself on a subpar performance. She was 64 years old, at the tail end of her bracket. But looming like Christmas was the world indoor championships in Finland. She would be 65 then. A blank slate awaited, and Raschker aimed to write a whole passel of new records on it.

Much depends, then, on managing our expectations for ourselves as we age. Olga's own expectation management system is pure genius.

She has two modes of thinking about the events in her life: one for things she can control, such as performance, and one for everything else. For things she can control her expectations are sky-high. For things she can't control, her expectations are modest.

Expectations and outcomes both affect the dopamine reward circuitry. Dopamine, remember, is the drug of desire and interest and focused energy—qualities almost synonymous with youthful purposefulness. Expectations of success cause dopamine to flow. So does unanticipated success. The big bringdown, biochemically—the thing to avoid at all costs—is *unmet expectations*. Unmet expectations, the University of Cambridge psychologist Wolfram Schultz has found, make dopamine levels crash.

Olga's strategy means she is pretty much constantly surfing on high dopamine levels. She is perpetually in a state of being either confidently anticipatory or pleasantly surprised. She is brimming with desire and interest.

I have asked dozens of category winners in masters events if it gives them pleasure, in their training, to beat younger athletes. Damn right it does. Especially younger athletes who know nothing about them and assume they must be slow because they're old. That's where the chip on the shoulder comes from.

On the alpine ski-racing tour, when a woman beats a man in a training run, that man has been "chicked." A humiliating beat-down it is perceived to be. (A tipping-point moment in a guy ski racer's career, one racer reported, is when you "no longer get beat by the best girl. When you are no longer getting 'chicked.'")

Both Olga and her friend Christa Bortignon enjoy chicking the guys they sometimes race. It doesn't happen much for Olga these days, but it still does for Christa. In fact, Christa's drive to beat the guys is a prevailing theme in her life.

She was the only girl in a family with three boys and a

mother who transparently preferred boys to girls. Christa was a gifted athlete, and at age 15 she asked her mother to sign the form so she could join the local track club. Her mother replied, "Girls don't do that." Christa took the house key and refused to give it back until her mother signed the form. Her mother relented. (But Christa would later learn that her mother had been approached by officials from the German Track and Field Federation who were scouting talented youth to develop for the 1956 Olympics. Her mother had turned them away.) At 20, Christa left the country with ten dollars in her pocket. In Canada, where she landed, she clawed her way up in the male-dominated field of accounting. She worked in Ottawa in a government division rife with chauvinism. She recalls a conversation between the deputy director and the director.

DD: "I see you hired a woman."

D: "Yeah, but I put her somewhere where she couldn't do any damage."

That memory is all the motivation Christa needs. That director is every guy on the track who believes she has no business being there—until he gets chicked.

Olga's drive to race the guys is different. She doesn't need to beat them. She just tries to stay with them. They are motivation to pull her along.

Olga's writer friend Roxanne Davies floats a superhero origin myth to explain why Olga strives the way she does. Olga has no birth certificate. Ninety-odd years ago, when a census taker showed up at the family farm in Saskatchewan, her distracted father miscounted the kids. He forgot about Olga. No birth certificate, no record of existence. In that moment a furious Olga Nobody was born, bent on spraying her name on the wall in indelible paint.

In Montreal, walking back to the hotel one day, I put to Olga my own guess about what motivates her. She grew up in the middle of eleven siblings, a lost sock in the drawer. Now she hears every possible variant of "Olga, you're one of a kind!"

"You're getting the only-child experience you never had," I said. "Now you can have all the attention."

"Well, that's what people say," Olga replied.

"You don't feel it?"

"Not really," she said. "I started playing Slo-Pitch at 70 and a new life opened up for me. Prior to that I was Plain Jane. And I'm *still* Plain Jane."

Olga's daughter Lynda Rabson, who brooks no BS, overheard my question and made a mental note to set me straight. Her mom was never lost among her siblings, Lynda said. She *did* have a niche. Graduating with a university degree while teaching all day and raising two daughters as a single mom? She was the enterprising one.

There is another theory of what makes Olga run—or, rather, want to run. It's not something she would necessarily recognize in herself because it is genome-deep.

In a spacious cage in a cramped lab in the psychology department at the University of California at Riverside, there lives a white albino lab mouse who has no name but should, so I will call him Dean. Dean is small—just three-fourths the size of most mice of his species, and very slender, with long, thin musculature. Dean has a caffeinated air to him; he's constantly twitching, almost never still.

Dean may be the world's fittest mouse. He has been bred to run, which is to say, he comes from a line of mice—sixty-four generations long and counting—who like to run fast and far. Most mice like to run. Some mice, like some of us, become bored with repetitive exercise, while others just keep on trucking. In this experiment, which began nearly twenty years ago, mice who ran the farthest after six days of trials were pulled out and bred with one another, producing supermice who ran still farther; that trait of running fast and far was amplified with each successive generation until you ended up with a mouse like Dean. When Dean wakes up in the evening (mice are nocturnal) he typically goes straight to his wheel, before eating, even, and just runs full-out, making the wheel squeal. He has run as much as nineteen miles in a day.

Many scientists are interested in how Dean and his siblings can log that kind of mileage. What physiological features are in place to support that level of activity? Turns out, neither Dean's heart nor his lungs is remarkable. What has evolved, in him, is a fantastically efficient system to deliver oxygen to his muscles.

Justin Rhodes, the experimental psychologist at the Beckman Institute who joined this study at generation twenty, is less interested in *how* Dean and his brethren run like that than *why*. Rhodes guessed Dean was getting a bigger psychological payoff than the control mice. Something in his reward circuit had been altered. He was getting juiced on exercise. And when you deprived him of his running fix he got extremely bummed. When Rhodes scanned his brain, he found high activation in the area associated with cravings for drugs such as cocaine.

Rhodes thinks genetic differences in the reward circuits may be one of the keys to the difference between why some people, such as those control mice, like a little exercise, and others, such as Dean, *love* and *need* a whole lot. But it's not quite that simple. You may have genes that predispose you to loving your workouts as much as Dean does, but genes themselves aren't enough. The opportunities must present themselves. Then the snowball starts to roll. The exercise feels uncommonly good, prompting more exercise, which delivers those bigger dose effects in mood and energy, and soon you are an exerciser. And if you have the right physiological makeup to support a high level of effort, soon you are an athlete. And now the intensity of the training starts paying big dividends in strength and stamina, and soon you are a great athlete. And if you keep it up, and avoid injury, you have bought yourself a life span five, six, seven years longer than the sedentary mouse in the next cubicle.

Dopamine operates by demand. Once you get it flowing, it releases enzymes that create new dopamine receptors. It's like Starbucks: as coffee sales rise, new stores magically open. Regular exercise keeps dopamine stores high. But it isn't the only thing that does. Studies show that winning produces a dopamine hit. So does approbation. When we do something very well and are recognized for it, the dopamine flows. In other words, it could be Olga's sense of accomplishment, as much as the exercise itself, that is so physiologically seductive.

Olga is a bit like Dean the mouse, Rhodes guesses, in that she just happens to have a suite of genes that, together with the environment she was raised in, "produced a brain that functions that way—that's highly motivated for physical activity."

You can see where this is going.

If there's a combination of genes involved in really, really liking it, maybe those genes can be artificially turned on in the rest of us. Not only would exercise become a compulsion, it could conceivably trump other unhealthy compulsions. Turns out if you feed Dean the equivalent of cheeseburgers, he doesn't get drowsy and lie down, as most of us would. He runs *more*. On some internal signal—possibly sensing high blood-leptin levels—he intuits the need to burn that crap off.

Pharmaceutical giants are surely wringing their hands in anticipation of synthesizing what has produced Dean the mouse. But before you invest in the company that's working on the motivation-to-exercise drug, know this: there is a Faustian hitch. "The more exercise the better" is true, but only up to a point. Then the curve turns the other way.

You wouldn't want to be like Dean. Running is fun for most

mice, but it's not "fun" for Dean. There is a horrible, insatiable desperation to his life.

"What these mice are doing, what they have become after sixty-four generations, is far beyond Olga," Rhodes says. Jonesing for dopamine, they are pushing themselves to the limit every time they run, and "they are actually damaging things. They have lots of new neurons growing but they don't show the cognitive benefits of normal mice that have been running. We don't know why that is. Part of it is that their dopamine circuits have been altered in some way. The symptoms are reminiscent of ADHD."

Of course, we may share much of their physiology, and even behave like them on bad days, but people are not mice. Beneath the pursuit of pleasure and the avoidance of pain, more profound motivations stir within us.

In the spring of 2012, when Olga and I had planned our visit to Illinois for brain testing, our schedule included a layover in Chicago on Saturday night. That left us twenty-four free hours in one of the world's great cities. "The day's your oyster!" I said. "How would you like to spend it?" Her answer, I reckoned, would tell me something about her strategies for squeezing maximum life out of this life. What inside of her was begging to be fed?

I threw a few options on the table.

The Botanic Garden?

Well, there are nice gardens in her own town, she said.

The Field Museum, with its famous T. rex skeleton?

Maybe if we had more time.

A Segway tour of Grant Park?

"What's a Segway tour?" she asked.

"You ride on these stand-up electric scooters," I explained. "George Bush fell off one but you wouldn't. You'd like it."

"I think you're confusing me with you," she said.

She *was* keen seeing a Cubs game, but the team happened to be on the road that day.

Then I remembered: it was a Sunday. Sunday is the fixed star in the rotating firmament of her week. She never trains on Sundays because she goes to church.

Chicago is home to St. Joseph the Betrothed Ukrainian Catholic Church—one of the more architecturally striking churches in the world. We investigated morning services. In the end she opted instead for the Art Institute, a more secular temple, which was closer to the hotel. She could miss a week's church attendance, she assured me, because she's already in pretty good standing with her god.

The issue of faith must be part of any analysis of what makes Olga run. On nights before competitions, Olga prays. It's always the same prayer—or rather a five-to-seven-minute-long sequence of prayers—and it involves a petition to stay safe and perform at a high level. What rescues this from narcissism is her sense of being an instrument of a higher purpose. One of Olga's favorite movie lines is from *Chariots of Fire,* where the Olympian Eric Liddell says, "God . . . made me fast. And when I run I feel His pleasure." Others of her generation, including Orville Rogers and Ugo Sansonetti, told me flat out they "run for God," though I've never heard Olga put it that singularly.

Whatever your position on faith, there's no denying it gets results. The faithful are onto something, in that they live longer and apparently happier lives, moment to moment. There's evidence that they use their time better and are better long-term planners.

Many elite masters athletes profess deep faith—it came up again and again in our discussions. Now, this is partly a generational story: most older North Americans were raised in a faith, which means if you're without belief at Olga's age you probably had to actively give it up at some point. But there are other ways in which the overlap between belief and high-level sports might not be coincidental.

Athletes get better by pushing themselves to their "ultimate limits." They emerge stronger than they would have been without that test, just as the repentant sinner is often said to be "closer to God" than the righteous soul who always stared straight ahead, and so never glimpsed the temptation that was shadowing him in his blind spot. Adaptation is a physiological fact, but it's also a spiritual notion.

There is a sect of Buddhist monks near Kyoto, Japan, who aim to magnify their suffering and ultimately transcend it by running. They run unbelievable distances—the equivalent of one thousand marathons spread over seven years, and they do it in sandals, on meager rations, through the mountains. The last year of this spiritual commitment is the toughest: two marathons a day. (The stakes are high: traditionally, monks who fail to meet the requirements are obliged to commit suicide.) The monks run not for their health or for vanity or for sponsorship dollars or for a runner's high—though spiritual transcendence sounds like the runner's high squared. They run for meaning. Religion spreads "a sacred canopy of meaning over chaotic lives," as one essayist put it, and pushing physical limits can bring exercise under that canopy.

But if faith works, what is the active ingredient that *makes* it work? What's going on when a believer lives two years more than a nonbeliever? Probably not the favor of some god with its thumb on the odometer to stop it from turning. No, it's the belief itself. The conviction. Conviction in something—an idea or a philosophy, such as social justice—can substitute for faith in a particular theological (or nontheological) story. The power of faith is really the power of purpose.

But purpose can be hard to manufacture in a vacuum.

The great American sprinter Michael Johnson was once asked if he had a prerace ritual, and he replied that he had only a *post*race ritual. "I stand on a podium and have them place a gold medal around my neck." That's pretty much Olga's routine, too. The difference is, where Johnson knew he had beaten people, Olga doesn't always have that satisfaction.

Frankly, having no one her age to race against is . . . getting old. At some point, knowing that you're going to walk away with a gold medal providing you cross the line upright starts to seem a bit ludicrous.

I could already see this wearing on her in Kamloops. Every afternoon Olga would arrive at the medals area and a couple of Mounties in their smart red serge and their wide-brimmed hats would escort her to the podium where she stood, alone. *"Too much,"* she'd say, under her breath. To cut the loneliness of that moment, she would often persuade the presenter to stand up there with her.

At the 2012 Canadian Open in St. John, New Brunswick, Olga arrived anticipating some real competition: a Mexican athlete named Modesta Martinez, new to Olga's category, was listed on the official roster. But Olga didn't see her anywhere, and she asked what was up. Turned out she had died a month earlier, and that information never got to the officials who made up the draw. "An oversight on their part, obviously," Olga said.

Around three years ago word emerged out of Japan of perhaps Olga's first rival on the planet. A woman named Mitsu Morita was burning up the track. Mitsu had become somewhat famous in Tokyo from a Nike print ad. It's an extreme close-up of Mitsu holding a shot put tight under her chin. Her skin has a kind of patina; she looks as if she herself is made of annealed steel. She claims to get her strength from eating eel.

Mitsu, who is about three years younger than Olga, is actually faster than Olga was at her age. Her world-record time in the 200-meter dash in the W85 category is almost ten seconds faster than Olga's current mark in the W90. Mitsu and Olga have never met and, since Mitsu doesn't compete internationally, it seems unlikely they will.

Why go on, then, running against ghosts?

Studies show that social pressures often prevent older athletes from quitting. Competitors carry on to avoid acquiring the label of "quitter," and feeling like one. Or else they're spurred on by the enthusiastic support of training partners and family. The single strongest source of positive social pressure, Bradley Young, a kinesiologist and sports psychologist at the University of Ottawa, found, was encouragement from a spouse.

Here Olga bucks the trend. She has no spouse—hasn't for half a century. John was a cruel drunk, and the best that can be said for their marriage, a decade-long trial, is that it finally ended. So Olga doesn't run because a spouse is encouraging her, and certainly not to get over the pain of his loss.

Not a few professional athletes—at least the more introspective ones—have tried to explain their own determination to succeed by tracing it to someone who very vocally doubted they could. That negative message, really the opposite of support, can nonetheless be its own kind of powerful fuel. Olga may insist today she has "nothing to prove." But that's only because, after a half century of ceaseless personal renovation, she has nothing *left* to prove.

"I think that has had a lot to do with her independence, with her drive, with her necessity to succeed," Lynda told me quietly one day as we sat together in the audience watching her mom give a talk. "As an abused person all you're told is that you're garbage. You're nothing to anybody. And so that made her say, 'You know, I can do this. And I can do it all by myself.'" If faith gives Olga the courage to run, faith in herself might just be what gives her the reason.

10

Olga and Me

I EARLIER REPORTED that it feels like Olga and I are going in different directions. Now let me explain why. Because I do think there's a good chance the two trends are related.

I am a man in midlife—a stage Olga was done with so long ago it's but a dim memory for her. In midlife, when there are health issues, it's hard to tease out the root cause. You are distracted and cash-strapped, and if you married late you are doing it all with kids hanging off you like koala bears. Is something actually *wrong* wrong, or is it just stress and exhaustion talking? It seems very likely that those telomeres would stop burning down like sparklers, and color would return to your face, if you could just string together a couple of good nights' sleep.

Even against that backdrop, though, some genuinely troubling signs began emerging around the time I turned forty.

Short-term memory went AWOL. I nearly incinerated my mother-in-law's apartment building after going out and leaving a baby bottle sterilizing on the stove. "Presenile dementia!" friends would chucklingly diagnose as I misplaced my wallet or keys or glasses yet again. The half-life of important details seemed to be about five seconds. If I didn't audibly repeat them they were gone. "Coffee cup's on the roof," I'd chant, buckling a kid into the car seat. "Coffee cup's on the roof." "You're only getting around fifty percent of what's going on!" my wife, Jen, muttered, exasperated, one day—which stung, although I had to admire the precision of the estimate.

The night before an early scheduled research trip with Olga I rustled up my passport, only to discover it had expired. It had slipped my mind to renew it. The trip was off. Everyone involved at the other end—the psychologists and scientists and lab techs who had tweaked their schedules to accommodate us—would have to be told. But first, Olga.

I called her at 7:15 a.m. with the news.

"Of all things to forget about—the passport!" I said sheepishly. "I must be losing my mind."

"Well," she said, sympathetically, "you haven't been using it much." (I think she meant the passport.)

It seemed unnaturally swift, this descent into low functioning—especially if Olga is the yardstick. Not only has she not missed an appointment in the years I've known her, she has never even been five minutes late.

Meanwhile, other signs of accelerated aging emerged. A potbelly seemed to just hatch. When I lay on the bed to read to the kids and rested the book on the waistband of my pajamas, half the page disappeared. Sometimes 5-year-old Lila would pull back, breaking the lovely Tupperware seal between us, and make a face. I smelled different, she said. Worse.

Bags appeared under my eyes. Senses became dulled. I could not read the serial number on the iPhone no matter how close I held it to my face. Could not hear it, half the time, when it rang in my pants.

One summer Saturday my pal Mike and I took our kids to the fair. In a back eddy of the midway, a young woman sat in a little booth. She was dressed in black yoga attire: a New Age gypsy. A sign said: "I guess your age within three years or you win."

I pressed five dollars into her hand.

She sized me up, top to bottom. She scribbled a number down on her notepad. She looked extremely confident. She had not missed all day, she said.

"What have you got?" I said.

She showed me the number.

"Doesn't really feel like a win, does it?" Mike said moments later, as I stuffed my prize, a giant teddy bear, into the backpack. What was there to say? This woman, an expert in decoding all the little tells that reveal a person's true age, had me nosing toward retirement.

That night I read the kids a story about a kindly man who worked in a shoelace factory. He had thick glasses and a fringe of graying hair. "Am I going to look like that in ten years?" I asked my 9-year-old, still thinking about the gypsy's guess. "Probably," she replied. "If you're still alive."

Now, this was all just too weird. For most of my life I had looked *underage*. I was carded in bars into my 30s. Then wham. Some kind of switch got thrown—exactly the opposite of the switch thrown in Olga.

"Tell me your symptoms," my family doctor, Rocky, said as I settled into his examining chair.

"Well, for starters, I basically can't run anymore," I told him. "The springs are shot." You really couldn't call it running anymore, I said. It's jogging. And even that's being charitable. I'm sucking wind out there. Many days I'm reduced to the dreaded walk/jog: an old guy's ritual. And any kind of slope at all does me in.

Anything else? Rocky asked.

Yes. I toggle back and forth between sleepy and dizzy. The sleepiness I blast with caffeine and the dizziness with trying to remember to breathe. And I'm just so weary. Weary on what seems like a cellular level. I feel like a goldfish owned by a kid who never changes the water, ever. If there is any testosterone left in me, I doubt Lassie could sniff it out.

Sure, sure. *Some* slippage is to be expected. "Our overall health naturally declines at around half a percentage point per year, owing to normal cellular aging," notes Walter Bortz, the Stanford medical professor and past chair of the American Geriatric Society in his book *Next Medicine: The Science of Civics and Health*. "So if you notice a decline sharper than that, something else is going on."

"I think something else is going on," I told Rocky.

Not long ago I made a list of things that have been part of my life but not Olga's—from dinners microwaved in plastic Safeway bags, to so much time spent slouched in front of the TV that I can still recite McDonald's commercials *backward*, to mercury handled in science class, to fire retardant in the furniture, to a thousand Cokes served with ice dirtier (one study found) than the water in the restaurant's toilets. The only reason I didn't also frolic in DDT mist is that the spray trucks never made it down our street. Modern consumer culture, even if it deploys no one obvious assassin, harbors a lot of suspects, from trace chemicals in the water to microwaves in the air to whatever the livestock we just ate, ate.

Olga grew up on a farm, which gave her a head start, healthwise. It's easy to romanticize farm life as somehow more "natural" and therefore healthful than urban life, but U.S. Census data indeed reveal those who grew up on a farm in the American West—a shrinking tribe—are more likely to become centenarians than those who didn't.

Many things on my list belong to a world Olga has managed to be in, but not *of*. A lot of senior athletes of more or less her vintage tell of similar beginnings, good habits established before the seductive conveniences of postwar life. "I lived on a bicycle," says marathoner Ed Whitlock, of growing up in war-time London. "I'd use it to cycle one hundred miles a day. Because even if you had a car, then, there wasn't the petrol." His peer Earl Fee concurs. "For me, growing up, there really wasn't a lot else to do *besides* running. There was no TV and my parents didn't have a car, so the kids walked everywhere." He still does.

The differences between Olga's list and mine add fuel to a kind of "cumulative insult" theory, which is much on the minds of a generation looking for an explanation for their midlife malaise. It seems we may never really have been "healthy," even if we fancied ourselves "fit." And the bill for confusing those two words is coming due. Perhaps what we are now is *impaired*—by small doses of toxins too new to ban on the weight of evidence; by habits we thought were good but turned out to be bad, or habits we knew were bad but couldn't be bothered to fix, or bad habits we never questioned. In twenty years? We'll still be around; we just won't feel like *us*. This is the ticking doomsday clock of the boomer. But unlike history's previous casualties of epidemics they didn't see coming, we have, in theory, the knowledge to rescue ourselves, one good habit at a time. The Olgas of the world offer the way.

I have known Olga for only four years. But some days I think she has always quietly been there, nearby but invisible, like the water table.

In Canada during the 1970s, junior high school kids were subjected to a fitness program called ParticipACTION. It was a campaign born of shame, after news emerged that the average North American kid was less fit than a 65-year-old Swede. Olga was in her 60s at the time. We were measuring ourselves against old European stock. Olga, or someone very like her, was the phantom benchmark even then.

All these years later, as a grown man, I see her almost as a living embodiment of my father.

He was calm, as she is, and just as kind. Like Olga's, his athletic career flowered late in life. Once the business of raising four kids was behind him, he became a multisport jock at the highest level. He had preternatural coordination and endurance. At age 65 he won the World Masters badminton championship with such ageless grace that the guy he beat in the final made a formal appeal to see his birth certificate.

A year later he was dead from leukemia.

It made no sense. With bulletproof genes for longevity (his mom and dad lived to 101 and 97) and a blameless lifestyle, he should be alive today and, it's easy to imagine, almost as robust as Olga. If he were, he would be very close to her age. Maybe it's because I still miss him so much that I see him in her. She continues the story that he began.

"You've inspired me to try something," I told Olga when we met for a meal one spring day in 2011.

"Oh?"

The outdoor world championships were coming up in Sacramento, with competitors from eighty countries already committed. It was the most important meet in masters track and field, and certainly Olga's biggest event of the year.

"I just signed up for the 10,000-meter," I said.

"To cover it?" she asked.

"To run it."

I had a ready answer if she asked me why.

Partly, I wanted to go deeper into the reporting, to get a better feel for what she experiences at these meets.

Partly, I wanted to know where *I* stood. If I was going to try to "fix" my broken self by becoming a little more Olga-like— through the systematic addition of one thing at a time from her list (or, conversely, the removal of one dubious habit from mine)—I needed a baseline measurement. How did I stack up to masters athletes at the top of their game?

And, finally, because something I'd just read had jarred me deeply. It was a remark by the Nobel laureate James Watson, codiscoverer of DNA, concerning guys my age. "Men of fifty don't like to fail," he said, "which is why they are so dull." Watson was arguing that research grants ought to go not to the established old lions but to young scientists more likely to take risks. The point obviously, chillingly, extends. Fear of failure is such a negative trait that it ages people, figuratively, maybe even literally. We worry about botching the things we used to be able to pull off with ease, and so we stop putting our whole heart into life. That looks to the world like disinterest— "dullness"—but is, in fact, a sadder condition. The moment halfheartedness becomes a habit, something dies in us. Olga doesn't fail much, but her willingness to keep putting herself in positions where she *could* fall flat, very publicly, is one of the biggest differences between us.

I had these answers ready to go. But Olga never asked why I was doing it.

"Outstanding," she said simply.

Enter Through the Gift Shop

"This is gonna blow you away," promised a man standing beside a display of plastic wristbands. He gave me a shove. I had to stutter-step to recover my balance. He then fastened around my wrist a "holographic bracelet" sending out "frequencies highly compatible with humans and animals on a cellular level," according to a little sign. He pushed again, obviously much more softly, and pretended to marvel as I didn't fall over. "See? That's why three hundred athletes bought one of these bracelets today alone."

To get in and out of the main stadium at Cal State Sacramento, home of the 2011 World Masters Athletics Championships, the biggest masters track-and-field meet ever staged, you had to pass through the marketing tent.

The products moving well were designed to help seniors run fast and perform optimally. The products moving *really* well were designed to help seniors recover something of their old mobility mojo in everyday life—"balance, strength, flexibility and range of motion." If a little magical thinking salts the broth, so be it.

You could open your own alternative drugstore with the elixirs people were talking up: beet juice, pickle juice, cranberry juice, hornet juice, caterpillar fungus tea, extract of the astragalus plant. I purchased a pair of Incredisocks, made of bamboo-, charcoal-, and germanium-infused thread and guaranteed to "boost my blood circulation from between 20 and 50 percent." I was going to need all the help I could get.

Olga had already arrived and checked in. She had entered her usual eleven events. Me, just the one.

The 10K is the flag of the Everyman. Most recreational runners who are training for something are training for the 10K. The marathon may have the cachet, but of fifty million or so North Americans who run, only around one in ten will try a marathon in their lifetime. For a lot of us midlifers, running a 10K, then running another one, with luck a little faster this time, as the pounds melt away and the sap rises, is the Holy Grail. It's the Holy Grail because, increasingly, it feels unreal, unattainable—especially the "doing it faster" part.

I had been training, but nowhere near tenaciously enough to be competitive. I just didn't have the wheels. As hard as I tried, I couldn't seem to improve. Steve Prefontaine used to say he entered races not to see who was the fastest; he entered them to see who had the most guts. I've always wondered how he could tell. You never know what private battles people are fighting.

At masters meets, they make you run with people your birth certificate age, not your biological age. So I had to cook up an experiment to figure out where I really belonged. I would run my race—Men's 45–50—and then compare my time with the winners in the older brackets. How many age groups up would I have to go before my time would be good enough to win? Five? Six? Seven? Eight? I had a secret goal: to (virtually) win the Men's 80–84. To be the fastest damn 80-year-old in the world over 10,000 meters.

The next morning, on the lawn outside the dorm, a group of a hundred or so high school cheerleaders were working on a complicated routine involving high kicks and contortions that caused their short skirts to hike and spin. This arrested the attention of a group of athletes from Central America, who were making their way to the cafeteria for breakfast. The men, who appeared to be in their 70s, stood transfixed for at least a minute. Then they exploded into wild applause.

The cafeteria served the kind of fare that got me through college: cubes of unripe melon, treacly juice, pancakes that could stop a bullet. I noticed a slender guy with cropped, silvering hair who was taking a while to choose. He finally settled on a yogurt and plunked it on his tray wearily, like Mary Richards tossing meat into her shopping cart on the *Mary Tyler Moore* show. I thought he was American but was quickly proven wrong when he joined me at my table and started to chat.

His name was Michael Barrand, a middle-distance runner from South Australia, competing in a division two brackets up from me, in the Men's 55–59. He was 59, which meant either he was here for the fun of it, or he was a study in optimism, or he was really, really good.

Turned out Michael had come to this meet via Oregon,

where he'd been visiting his stateside "family." In 1969 they had billeted him, as a high school exchange student, for a memorable, post–Summer of Love sojourn in America. With some good coaching, he caught the running bug that year, and for the first time started taking track seriously. But upon his return to Oz, "life got in the way of running." Three decades later he resumed his training with a vengeance—as a volunteer surfside lifeguard, running in the sand.

Michael teaches history to middle school students and has a way of expressing himself that calls to mind the phrase "Ask him the time and he'll tell you how to build a watch," but in a good way. He is sardonic company. We talked of overrated writers and underrated runners. We loitered at the intersection of language and sports. We were still sitting at the table when a guy in an apron hustled us out of the empty cafeteria. Michael mentioned he was going to do a little geocaching later on campus and asked if I would like to come along. I told him Olga and I were having dinner that night and invited him to make it three.

It occurred to me that I had not actually made a new friend— that is, a friend whom I chose, and who chose me back, rather than a parent of someone my daughters had befriended—in twenty years. That's midlife for you. Your social net shrinks as your investment in your family deepens. Friend making is a muscle you lose, and when you test it again after that long while, some feelings come back that ran through you most strongly in childhood, like vulnerability and promise.

My race happened to be one of the last scheduled, so I had time to do some research. I sought out the winners of the other 10K divisions, as their races finished. I was curious what they had in common. Was there a story that would resonate with Olga's?

The winners came from four continents and all walks of life. (Though mostly from the middle class: you need some means to go out of pocket for the international travel.) All of them looked young. All of them said they *felt* young—twenty-five years younger, on average, than their actual age. All admitted they got a huge charge out of beating younger runners in practice. But something else emerged from their stories. Something important.

The Secret of Resilience

Heimo Kärkkäinen, 60, a bearded and soft-spoken Finn who works as an accountant for a health insurer, played soccer as a kid, and because he was too skinny for even the incidental contact of that sport, he turned to running instead. By his mid-twenties he was an elite marathoner. He clocked a personal best of 2:25. But he gave up marathons when he found he could no longer break three hours—because three hours is too long for anyone to be running at a stretch, he strongly feels. The past couple of decades he has throttled way back.

Emilio de la Cámara, 66, used to train with a kid named Rafael Nadal, a tennis player some thought might go places, in his native Majorca. Emilio is a former soccer umpire who used to put in hellacious road miles. He ran the Budapest marathon in 2:33 at age 45. Then he, too, stopped running marathons. Only lately has he returned to the track with the old intensity.

Peter Sandery, 70, a former school superintendent from Adelaide, South Australia, was always pretty sporty—he roller-skated and speed-skated and played some basketball and even shot pistols. He took up running at age 51 after a job change put him on the road a lot, since running is something he could do anywhere.

Bernardino Pereira of Portugal, 75, who looks like Silvio Berlusconi and runs like Rocky Balboa, took up running at age 64 after he developed varicose veins in his sedentary bank job and his doctor told him to get moving. You could deliver a newspaper through the crack in his grin. "This man is now happy," said his interpreter.

You see the pattern. None of these athletes, the elite of their class at this stage, put pedal to metal the whole way. They either started late or at some point "life got in the way of running" and they took a decades-long hiatus.

"Most of the athletes you see here started when they were forty or older," a British field judge named Mick Fraser told me one day in Sacramento. "Because if they started in their twenties they end up broken. Or else they end up like me, as officials." Fraser was himself a middle-distance runner until, by 50, he couldn't bear the pain in his knees. When he went to the doctor who diagnosed arthritis, the doctor told him: "I've seen people like you. They're like two people. From the waist up they're golden. From the waist down they're a train wreck."

Turns out, you find the same story across all sports.

In 2012 a French cyclist named Robert Marchand set a world speed record for his age—100—before a cheering crowd at a velodrome in Aigle, Switzerland; he took up the sport competitively at age 78.

Margot Bates, of Adelaide, who was still famously swimming competitively at 103, started at age 87. "Banana" George Blair, legendary nonagenarian barefoot water-skier, was in early on his sport, relatively speaking. He first tried it at 46.

And Olga, of course, took up track at age 77. The real surprise would be if she *wasn't* a latecomer.

The body should, in theory, be able to absorb a lifetime of punishment, so long as technique is good. (Running particularly, with its simple back-and-forth motion, should be easier on us than sports that make us plant and twist—or worse, plant and twist and get hit.) But let's face it: most people's technique is not good. And as we age and fall into bad habits, it gets worse. Competitive athletes overtrain and acquire muscular imbalances, and their limbs start to move outside of their assigned tracks, destroying cartilage and overstressing tendons. One of the secrets of masters athletes like Olga may be that both their heart and their chassis got a good long rest from high performance.

But it can't just come down to the body's engineering. A lot of this is surely about will.

Older masters athletes (say, 70 on up), the ones who end up at meets like this, are usually thrilled to be competing because this is all an exciting new chapter for them. After a midlife spent providing for the family, they finally got the chance to use their bodies for fun, not work, and they seized it. Forget that there's no prize money, no fame, not even any sponsors supplying free shoes; just getting to do it is payoff enough.

This may all explain why people who come to masters track meets to stargaze are sometimes a little disappointed. Where are the legends of the 1960s and '70s—the Lasse Viréns and the Frank Shorters and the Rosa Motas? Aren't these folks now masters? If they had run into the post-Olympic sunset and then privately continued on that trajectory, performing at that same high level, surfing the leading edge of their physiological decline, they would be here, driving the world records well beyond what they are. Shouldn't *they* be the Ed Whitlocks and the Olga Kotelkos of the modern age?

I had secretly hoped to spy one of my heroes, Carlos Lopes, the great Portuguese distance runner who did all old guys proud by winning the Olympic marathon at age 37. Lopes vanished into the "Where are they now?" file, only to resurface briefly on a *Simpsons* episode. (Homer, watching Olympic reruns, does the math and realizes he's Lopes's age, and is inspired to don a headband and nipple tape and enter the Springfield marathon, with predictable results. Marge: "Hey, Grandpa's running!" Lisa: "That's not Grandpa. Dad's just dehydrated.")

"Where's Carlos Lopes?" I asked the Portuguese team manager, Luis. He mimed a Buddha belly. "Injured?" I asked. He shook his head. "Money?" He nodded: "It's possible." Luis suggested a simple cost-benefit analysis was in play: Lopes *could* whip himself back into shape to be the best all over again, but why would he? One imagines the postretirement life of legendary runners as pleasantly senatorial: a taskless thanks. They bask in their memories, jog with the dogs in the dunes, perhaps sire some Kardashians—with no hunger to return to that old life.

But that's not always how it goes.

There had been a buzz, as Sacramento approached, that Henry Rono, the great Kenyan distance runner, might race. Rono's post-Olympic life is a sad story. After a slide into alcoholism and even homelessness, he'd ballooned to 230 pounds and found himself too broke to travel. Had Rono turned things around? Response in the community was guarded. "If he shows up out of shape it is almost worse for the sport," posted one observer on the masterstrack.com blog. Such potential embarrassments could be gracefully skirted, many seemed to feel, if there were mandatory qualifying times every athlete had to meet with actual recent-ish verified race results. To me that seemed against the spirit of the thing. But, of course, I'm biased. Such a system would have weeded me out.

With my race still two days away I bused across town to Hughes Stadium, where the final of the 10K for men aged 80 to 85 was just beginning.

A lot of senior tracksters run as if no one taught them how. But one entrant this morning had beautiful form. In an age bracket I call "the shuffle threshold," this runner gobbled up the track in long, efficient, effortless strides.

Ed Whitlock holds the distinction of beating the minimum

Boston Marathon qualifying time by the largest margin ever. The top 20-year-old Kenyans are besting their minimums by a little over an hour. Ed, when he ran his 3:15 marathon at age 80, was under by an hour and three quarters. (The Boston Marathon qualifying time for everyone over 80 is five hours flat. So the new bragging point for aging runners will be to become the oldest person to qualify for Boston—which means you're the oldest person to run a sub-five-hour marathon.)

Ed sliced through a wicked, swirling wind, his pace metronomic, round and round the track. He finished in forty-two minutes and change, shaving more than two minutes off the world record. If it's possible to feel both elevated and dispirited by an athletic event, that's what I felt. Elevated because, hey, it's a privilege to see such a commanding performance by an athlete of any age. Dispirited because I had vowed to "win" this age group. And I was pretty sure I couldn't run faster than Ed just did.

(As for my hiatus theory, Ed follows the script. He was a competitive miler in college. Then professional and family obligations ate up the whole middle of his life. He didn't run seriously for three decades.)

Olga, meanwhile, was having a good meet. She had run her 100-meters in 27.5 seconds. She seemed pleased with that, even though her record is 23.95. With Zola Budd, the former world cross-country champion and meet headliner, having pulled out with an injury and jetted back to South Africa, Olga had become the feel-good story of the games. After her world-record hammer throw, the media center issued a press release about the continuing feats of the "ageless British Columbian."

"I Was Just Trying to Distract
You from the Pain"

"Gentlemen, run a good race," the director said. "Start time is in six and a half minutes."

We were mustered in a little holding area beside the track. Runners paced, pawed the ground, affixed their race number to their hips and their hearts. There were all body types. A couple of Colombian gentlemen, happily, had guts bigger than mine. "You look fast," I said to one tall, svelte guy who resembled Timothy Hutton. "Looks are deceiving," he said, cryptically.

Over the previous twenty-four hours I had tried to tap the hive mind, all these decades of experience. Were there tricks to preparing for a race like this?

The studs swore by their various rituals. Gary Stenlund had his "fountain of youth" exercises, Earl Fee his quirky dietary strategies (no flesh and starch at the same time), Olga the business with her legs against the wall ("gets the blood to your heart and picks you right up"). Everyone was concerned about the heat, which had been climbing into the triple digits. One British athlete swore by holding ice in your hands to lower your core temperature—a technique he described as "better than steroids."

"Listen to everyone's advice—then feel free not to take it," said Ed Whitlock.

"Congratulate yourself for being here," said Olga.

"The big thing is," said Canadian 800-meter runner Bill McNamara, "you just don't want to come in last."

Alone in my maple leaf, I scanned the other guys' racing bibs. I'd written some numbers on my wrist. These were the runners who, based on their estimated finish times, I thought I could keep up with. I was looking for number 910. And there he was: big guy, perfect to draft behind. He kept jumping in the air. He wore a little Breathe Right strip on his nose. *Breathe*, I reminded myself. Breathe.

Our group was sent to the starting area to "stride out." I presumed that meant we'd take a little warm-up jog. I tried "striding out" for ten meters or so and then stopped, only to be crashed into by the guy striding out behind me.

Runners lined up elbow-to-elbow, and twenty-five laps ahead of us, and a sound like pistols at dawn.

No matter how many times people tell you not to go out too quickly, it's hard to listen. There's an irresistible urge to stay with the pack. *If I fall behind,* a primitive part of the brain understands, *I'm lion food.*

I was clearly carrying too much weight. I'd promised myself I'd shed ten pounds for this race. That would shave three minutes off my 10K time, by Earl Fee's calculations. But I'd made the pledge within reach of a bag of sour cream and onion chips at two in the morning, and my priorities became confused.

The pack hit the quarter mark of lap one at 1:35; the half at 3:10; the mile at around 6:20. Way too fast for me. The next mile was 6:30. I tried to hang on but couldn't.

There's a moment in novelist Haruki Murakami's running memoir when he realizes he hasn't left himself enough time to train properly for a race. You can get away with minimal training when you're young and still muster a heroic performance on game day, he concludes. But by the time you're crowding 50, "you really do get only what you pay for."

The lead runner lapped me, passing on the left. By now I was mostly just in people's way. At one point I felt a pain in my groin, on the right side, which I thought was maybe my vasectomy coming undone. I wanted badly to stop—and if it hadn't been for some voices of encouragement dopplering in every time I passed the grandstand, I'm sure I would have. One of the voices was Olga's. One was Ed's.

Michael Barrand, my new Aussie pal, had unexpectedly got knocked out of his M55 semi, so he was free to come watch my race. He parked himself beside the track and issued crisp little coaching tips each lap. "Shoulders back." "Keep to the inside." "Relax the top half of yourself." "Use your arms."

By the last mile I'd slowed to an almost nine-minute-mile pace. The bib numbers on my wrist list were all long out of sight.

Snowshoeing down the final straightaway I could feel my legs starting to buckle, and at the finish line they did. I rolled onto the grass by the scoreboard. It felt good to lie there. The concerned voice I heard next was an official. "Try to get up and move if you can," he said. "Otherwise you'll seize."

There is another level of embarrassed when you are representing your country and you run like a hairy goat. From how bad this felt, it was easy to infer how great it must feel for Olga. You're flying the nation's colors and you cruise to victory. The difference is as vast as the space between shame and pride.

Michael clamped a hand around my shoulder. He knew I was cooked, he said, when I started looking down at the track.

"What you wanted to do was keep your head up and look at the guy in front of you, at the backs of his shoulders," he said. "There's an imaginary line connecting you. Then very slowly you reel him in. When you're right behind him you can stay there for a moment, and then you just nip by him. And at that point the power balance shifts and he thinks, I can't stay with this guy, and *he* looks at the track and *he's* done. And none of that really takes much more effort than putting your head down in the first place. So that's what I was trying to get you to do."

"I could barely stay on my feet," I said.

"I know," Michael said. "Mostly, I was just trying to distract you from the pain."

My quads were seizing.

"Breathe," Olga said. She and Michael and Bill McNamara escorted me across the darkened campus and back to the dorm, guiding me by the elbows the way you would a blackout-drunk friend. Michael lent me his compression tights to sleep in, to speed muscle recovery. He handed them to me with quiet ceremony, like the invisibility cloak bequeathed to Harry Potter by his father.

The winner of my age category was a 47-year-old Spaniard named Francisco Fontaneda, who works as a purchaser for a retail chain in Bilbao. (Hiatus theory? At age 26, he quit the pro circuit and didn't run for nineteen years.) He finished in 31:32—a minute and a half off the world record. I thought about sending flowers to Gerardo Ramirez of the Dominican Republic. Because of you, sir, I did not come in dead last. But at 45:40 I was second last.

I averaged 7:49-minute miles. A guy has run faster with a milk bottle on his head.

Turns out I'd underestimated the number of age categories I'd have to climb before I would have won. The answer is *nine*. I would have won the Men's 85–89.

But, you know, by quite a comfortable margin.

The Payoff

As the midday sun hammers Cal State stadium, I take a seat to catch the start of Olga's 200-meter final. It's not her favorite event. And because all competitors 85 and up are grouped together, she is in tough company. Patricia "Pat" Peterson of Albany, New York, is in this race. She's the American record holder in the younger 85–90 age group and a big crowd favorite, partly for her inspirational backstory. Peterson conquered cancer three times. She is thought to be the oldest person to have received a stem-cell transplant. Just six months ago she was in bed rehabbing after slipping on ice and fracturing her pelvis—but you'd never know it from the way she surges out of the blocks.

The runners emerge out of the heat haze across the oval. Peterson is running neck and neck with Emilia Garcia De Fontan, a Colombian woman her age, with Olga just behind. The applause grows as the runners enter the straightaway and will themselves toward the finish line. It crescendos as Peterson edges Garcia De Fontan at the tape, then spikes again as Olga crosses, five seconds off her world-record time of 56:46.

And then something strange happens. The stadium falls quiet, and a trickle of clapping starts up again. People have realized there is still a runner on the track. Way back, still rounding the corner, tiny and bespectacled and wearing a brilliant blue head scarf. This is Mann Kaur of India, age 95. She has flown here to compete with her son, sprinter Gurdev Singh. It was her first time on an airplane, the newspapers are reporting. She is moving very slowly.

The applause grows. It changes. It becomes a rhythmic drumbeat—CLAP, CLAP, CLAP—the call of the tribe. Every iota of the crowd's attention is on this little woman. It takes her well over two minutes to complete the 200 meters. The cadence of the beat increases. The crowd roars as she crosses the line. Her family spills onto the track to meet her. They drape her in her country's flag. Everything is on hold. Goodwill fills the air. Eventually people start looking at their watches. The runners queued up for the next race pace like cattle. Mann Kaur seems to be about to leave the track and then unexpectedly turns back toward it.

"Oh, God," one of the Aussie contingent says. "I think she's going to do a victory lap."

There's only one place you'd get away with a line like that, where it's funny and not cruel: within a family.

And here is where I finally start to get it.

You can float a million armchair theories of what motivates Olga to keep doing what she's doing. But if you ask *her,* her number one answer is always the same. It's not the travel. It's not the prizes. It's not the fame. It's not the satisfaction of getting to occasionally whip the bony behinds of the men. "It's the camaraderie."

The twenty-first century belongs to friendships—virtual, yes, but real ones, too. That goes double for old people in the West, who are so often plunked on an ice floe and shoved out to sea. If they're lucky, something like masters track puts them back in a community—a purposeful and healthy place to be.

One of the most rock-solid findings in gerontology is that strong social ties boost your likelihood of surviving, over a given time period in late life, by 50 percent. The effect is larger than the impact of exercise. It's roughly the same as quitting smoking. There's evidence that cancer progresses more slowly in people with friends than in people who feel lonely. Strong social ties also correlate very strongly with healthy cognition—though it'd be hard to prove they *cause* it. To really lock the case would require randomized trials. "But you can't assign someone to have a big family," says Art Kramer, and most of us would prefer to choose our friends.

Friendships don't preclude rivalry, of course, nor is the converse true. Two opposing hockey players can punch each other's lights out and then go for chicken wings after the game. But the friendship/rivalry matrix of older people is different. Many older track athletes are no less competitive than younger ones. (Competitiveness isn't just hormones talking: it's temperament, it's *sisu,* and these are things age doesn't really change.) But increasingly, as we get older, competitiveness can coexist with friendship. Friendship/rivalry becomes less like a switch and more like a braid. Michael Barrand and his cohorts from Oz would call it "mateship." It's a kind of supercharged friendship, requiring more of a commitment—to the memory of how it began, to keeping the little flame of the thing alive long-term, to the ongoing well-being of the other person. "Each assists the other in living a good life, a life of virtue and happiness"— that's not Michael, but Plato, who himself spent a certain amount of time in the gym. If athletes grow into particularly rich friendships with one another over time, it may be that dimension of mutual improvement that explains it. They get better, and are better, together, and they know it.

In the 400-meter final for men 80 and up, the reigning world-record holder, Earl Fee, was overtaken down the stretch by a Japanese fireplug named Hiroo Tanaka, a new entrant to the category. In the finish area, Fee was steaming. Tanaka came over to him. He bowed and said something in Japanese. A translator tried to convey the exchange. Tanaka was trying to thank Fee for pushing him. But no, the idea was larger than that. He was giving Fee co-ownership of the win. "You did this," the translator said, looking at Fee. An ultracompetitive man, Fee was briefly struck dumb. Fee took Tanaka's hands in his, both at once. "You're amazing," he said.

That this is all terribly *hard work* seems to matter. After all, you can be in a book club or a coffee club with somebody for years and never really feel close to them. Comfort doesn't promote togetherness. Discomfort does. Nietzsche, who was never known to have turned heads in track pants, may nonetheless have said it best: "Exhaustion is the shortest way to equality and fraternity."

Suffering is the very name of the finishing weekend of a big masters meet. Competitors limp toward the massage tables. Some wander around with colorful strips of Kinesio tape on their quads and calves, looking like Illustrated Man and Woman, with their arteries on the outside. Masters athletes are like army buddies forever yoked by what they've been through—that minor (and sometimes not so minor) trauma. When you're 25 years old and watch a comrade trip on a hurdle and splat on the track, you may feel a tug of empathy, but there's no wholesale concern. When you're 90 and that happens, another part of the brain would seem to be activated—the machinery of genuine compassion.

In Olga's 100-meter-dash final at the world indoor championships in Kamloops—the one where she ran with the guys—Emiel Pauwels, the Belgian, caught a spike and crashed down, and Olga glided past him. But the moment she breasted the tape she circled around to face him as he limped in. They don't even speak the same language. But they didn't need to. It was like the Corleone brothers, a blood-deep understanding: *Emiel, this is the life we have chosen for ourselves.*

In Sacramento, Olga had been making a real effort to introduce me to her friends from back home. "Have you met Myrtle and Jerry?" she would say, as if confident we would make a good bridge foursome, maybe take in some shows. They are in their 80s. Everyone in my dorm room was over 70. But by the end of the week, they didn't seem old. (Well, one Slovakian gentleman sort of did.) This week these were my people—fellow travelers in whatever it was we were, we are, embarking on.

On Sunday, the athletes checked out and headed to the cafeteria for a last breakfast. E-mail addresses were exchanged. If you were looking for the headwaters of emotional commitment to the sport, you might look here. Why do this? It is a tautology. You do this to stay healthy enough that you can do this, thereby earning the right to reconvene with these people once again. To sit in the stands for their race and chant them home.

"See you in two years," I overheard one man say to another in parting, as he schlepped his friend's suitcase to the curb. "If you don't hear from me it means I died."

11

Going Deeper

IN THE EARLY 1900s a French doctor developed a specialty: transplanting monkey testicles into middle-aged men. It was meant to reinvigorate them. The doctor couldn't keep up with demand. Hundreds of men every day asked him for the procedure.

When he asked them why they felt they needed the energy boost, most men claimed they were eager to "complete some life work they had undertaken, or see through some particular enterprise that had not yet been crowned with success," reported the *New York Times*. That's how strong and deep is our desire to leave a positive legacy, to be remembered for some concrete accomplishment. Our failures mostly leave no trace.

Sporting marks endure. Whether world bests or mere personal bests, they exist, at the very least, as proof of what we were once able to do—of, in a sense, who we were. (How many times have you Googled someone from your past, and the only thing the search turns up is a marathon time? From it you can actually sleuth out quite a lot—roughly where they're living, how they're taking care of themselves, even what stage of family life they are in, judging by the training commitment their finishing time would have demanded.) When masters athletes grow too old to push their genes into the future, they can at least push their name into the future. It is a kind of immortality.

But I have often wondered whether, as a motivator, this legacy is enough. Aside from your determination level, leaving a high-water mark says nothing about your character. You can be a bad person and still be remembered by history (Ty Cobb is in Cooperstown). For a genuine sense of fulfillment, doesn't there need to be something more? Some dimension of service?

One day over lunch Olga and I were talking about a guy named Doug Eaton, also known as the "Reverse Panhandler of Oklahoma City." Eaton spent his 65th birthday on a downtown street corner holding up a sign. It read: "I have a home, and a car, and a job. Do you need a few bucks for some coffee?" He handed out five-dollar bills all day long.

Olga found the story of the reverse panhandler bemusing, and it is. But there is a seriousness to his efforts. Here was a man trying to make himself useful, at a time when the culture tells us we no longer are. To have a mission in life, something to get up for, some valuable role to play: that is a huge part of aging gracefully. The Japanese have a word for it: *ikigai* [eekee-GUY]. Rough translation: "the belief that one's life is worth living." Studies have found that those who have *ikigai* live longer. Some believe this is Darwin talking. Nicholas Humphrey, emeritus professor of psychology at the London School of Economics, has argued that feeling "special," feeling we have a unique purpose, has an evolutionary source. If we feel special, we'll take pains to stay alive.

The purpose-driven life is qualitatively better, too. Finding a raison d'être makes us deeply happy, many studies show. Indeed, one of the few things as strongly linked to lasting happiness as exercise is volunteering.

The impulse to be of service grows with age, some studies show. In a few seniors it's so strong that they will forgo comfortable retirement and focus instead on what they can still contribute.

"Have you heard of the Skilled Veterans Corps?" I asked Olga one day. This is a group of a couple of hundred Japanese seniors, many of them retired engineers or factory technicians, who have made themselves available as "first responders" in the event of a national disaster. If another Fukushima meltdown happens, for example, they have pledged that they will walk right into the radiation and work on the reactors to prevent further damage. Old people are uniquely suited for such duty, they believe, for while they may get deadly cancer, they'll probably die of natural causes first. They are willing to bear the hazards because they have already lived a full life.

"What do you think?" I asked Olga. Was this the way of the future? A certain breed of seniors leverages their benevolence by taking on these kinds of dangerous roles—in war field hospitals and on Mars missions and in nuclear hot zones, helping with the cleanup?

Olga thought about it for a long while. "I think it's very personal," she said. Service is good. But an obligation to serve? That she wasn't willing to concede. Because as soon as people feel obliged to do anything, then the act is coming from the wrong place.

Sport obviously provides Olga with *ikigai*—the sense of purpose in life, a sense of tapping the most uniquely valuable thing she has and sharing it. It's *possible* there are other world-beating skills latent within her—maybe fly tying or animal tracking— but it's safe to say that Olga found, albeit at the eleventh hour, the thing she was meant to do.

The question then becomes, how do Olga's private satisfactions translate into broader benefits? Just how powerful is the example of her life?

It depends on what we choose to make of it. If we decide we

want more people like Olga, then educators, psychologists, architects, and planners face the task of renovating the world— starting with figuring out ways to get more people moving. If everybody exercised, according to some estimates, the gains in health-care-cost savings would be the equivalent of the discovery of antibiotics. But most people won't exercise just because they're told they *should,* so the only solution is to remove the choice.

And so the design nudges commence. Cities become webbed with bike lanes that make getting around that way easier and faster than taking the car; public parks are equipped with seniors' "playgrounds," full of exercise and agility stations—a phenomenon increasingly common in Japan and some European countries. People are subtly persuaded to leg it out somewhere every day, until the habit becomes so much a part of daily life that we stop thinking of it as "exercise" at all.

Postretirement life is reconceived. Seniors' residences have plenty of open space and communal space, and few televisions. (You might even find, if NASA researcher Joan Vernikos has her way, the odd trampoline.) There are language and music programs, and the opportunity for new skills to cross-pollinate in aging brains. Residents can still retreat into private rooms, but there's much reason for them to venture out frequently: the choice is theirs. (So much of the quality of life boils down to "role and control," as the evolutionary biologist Michael Rose puts it.)

Staff have a background in both science and social work. They tailor diet and exercise programs to each resident's unique biochemistry, but they are also deeply *emotionally* invested in their clients. Studies show that level of care improves patients' memory performance and can even change the outcome of diseases.

Out in the community, great pains are taken to put the young and the old into contact with one another—something that benefits both sides, as long-lived populations such as the Japanese of Okinawa understand. Not long ago Anne Lamott mused that the question "How old are you?" might have an unusual correct answer: "Every age I have ever been." That may not fly on a passport application, but it's a useful thought experiment. We all need to be reminded who we once were (if we're old), and to imagine who we could yet be (if we're young). And maybe the best way is over a drink with that other generation in some sun-splashed piazza.

Or even over sandwiches at the West Vancouver Seniors' Centre, where Olga and I found ourselves lunching among the pensioners not long ago.

"Why are we still in each other's lives?" she asked, out of the blue, in her matter-of-fact way. "Why are you still following me around?" she unnecessarily clarified. "And why am I still letting you?"

"Well, I can only speak to the first part of that," I said. "I'm amazed by you, and I've grown close to you. You remind me of my dad." Yes: a mensch like him, a stud like him. The difference is, he died early, inexplicably and unfairly, and you endure, with near-miraculous staying power.

"In some ways your life bears out everything my dad believed but couldn't prove," I said. He was a psychologist—that was his job. But he was really a sportsman. That was his identity. He was all about mental health and physical health and he knew—he *knew*—that the two were linked. He thought fitness was the greatest antidepressant ever invented. "His life was a hymn to exercise and attitude, just like yours is."

I dipped my sandwich in my soup.

"And so," I asked her. "Why are you still letting me?"

"I guess I trust you."

That was probably true, but it wasn't the whole truth. The whole truth comes back, I think, to this dimension of service.

This is the "bigger purpose" that drives Olga's life now. "I am an open book," she is saying. Poke, prod, siphon, scan, interrogate, use me; just make it count. Make something meaningful from my strange uniqueness. Take away a few cells, a few traces. I can regenerate (I heal quickly). As the science inches forward, and new Olgas join the study, the picture will sharpen. Olga got this started, but she will never learn precisely what part she played. Her data will be meaningful in ways no one can predict. Thus, proverbially, do "old men plant trees whose shade they know they will never sit in." Old women, too.

At this stage of our understanding of aging and senescence, there remain, frustratingly, far more questions about what affects longevity than answers. Scientists have rounded up a vast pool of suspects but made few arrests.

It's frequently unclear whose side the suspects are on—the helpers' or the hinderers'. Supplements—of vitamins, minerals, hormones—are the ticket to a healthy, long life. Or else they're counterproductive, because taking supplements tends to cue the body to decrease its natural production of the very thing we're after.

Coffee keeps us young; it fights aging by stimulating muscles and preventing falls, and it reduces risk of stroke and respiratory illness. Or else it ages us because, as a stimulant, it disturbs sleep and increases blood pressure.

Sunshine shortens our life by boosting our risk of skin cancer. Or it lengthens it by lowering our risk of heart disease.

Cranky people live longer. Or sweethearts do.

Statins are the magic pill for those of us who need to get their cholesterol under control and stave off heart disease. But George Vaillant found with his Grant study of Harvard men, "The men's cholesterol levels at 50 had no bearing on how long they lived."

But Olga's test results, as they trickle in, have already added grist for further study.

The Beckman Institute team, in its analysis of Olga's brain scans, found much to celebrate. Her processing speed was better than most of the controls a quarter century younger. And there was a strong chatter through her corpus callosum—a dense fiber tract that runs from the front to the back of the head and connects the two hemispheres. (Which makes sense: for Olga to do all the different events she does, remembering and coordinating complex actions at that level, suggests some pretty sophisticated air-traffic control.)

But there were signs of aging, too. Above the right ventricle, quite visible in the images, Olga's white matter looked like a painter's drop cloth, embellished with little white spots. These were white-matter lesions. It means some roads are out. It'd be alarming to see so many lesions in a young person—"if she were 45 she'd be in an institution," said Art Kramer. But it's expected in people of retirement age or older.

A case can be made that Olga's general sharpness owes much to high activity levels over a whole lifetime. There was fantastic structural integrity in two places in the brain known to respond to exercise training. "If it were just genetic, it's hard to believe it would be just in these two regions," says Kramer. Overall, the results seemed to support what everybody suspected: if you want to keep your marbles, and you're planning to live into your 90s and beyond, start exercising now, if you don't already.

Likewise, early returns from some of the muscle tests suggest that exercise is helping slow the clock there, too. Olga has around eighty-five motor units still working to flex the muscles in her shins. That amounts to "the neuromuscular integrity of someone thirty years younger," the research team from Western Ontario concluded. With exercise, they surmise, Olga has been able to "slow the aging process substantially but not stop it."

The job remains to tease out precisely what's going on—how exercise is doing that regenerative work. It could be that the continual electrical stimulation of those motor units is keeping the circuits well oiled and functional. It could be that chemicals produced during exercise strengthen the nerves to the muscles.

More tantalizingly, it may turn out that the two most dramatic effects we now know about—that exercise protects the brain, and exercise protects body muscle—are somehow related. That is, the second is a trickle-down effect of the first. If true, it would amount to finding a kind of missing link between diseases such as Alzheimer's and diseases such as muscular dystrophy.

But other results hint at deeper mysteries. In the two years between tests, Olga lost virtually no wind and no strength—even though she had cut back her training. That seems to bear out Hepple's hunch that something else is going on in the muscle, irrespective of exercise effects.

Either way, Olga reckons she has never put her legs to such good use. Even if this time there is no gold medal to show for it.

12

Shadows

The O Word

IN THE SUMMER of 2012, during a routine medical exam, Olga's family doctor, Nigel Walton, caught something on an X-ray. A couple of vertebra in her lower back didn't look right.

Our skeletons are shockingly light, given the work they do, and it turns out Olga's is lighter than most. It weighs around 1.6 kilograms. That's about three and a half pounds, skull included. The *bacteria on her body* weigh that much. (My own skeleton, by contrast, is about twice as heavy as Olga's.) Her bone density is low in those lower vertebrae, and in a few other places.

No getting around it: she has osteoporosis.

It's a surprise, and it isn't.

"As we get older our bone density falls one to three percent a year," Dr. Kevin McLeod explained to Olga in his office a couple of weeks after Walton referred her to the specialist. "Most of us are going to have diminished bone density at ninety-three."

McLeod runs the osteoporosis clinic and cardio rehab program at Lions Gate Hospital. His job today, as Olga viewed it, was to help her decide what to do with this new information.

"I actually tried to look up average bone density for someone ninety-three years old, and there's no data," he said. He leaned closer. "We like to be evidence based in medicine," he said. "Ideally, I want to be able to look you in the eye, Olga, and say, there are a hundred people like you that we studied and this is the right answer. But there aren't. Most of our studies end at ninety. A lot of them end at eighty."

McLeod allowed himself a digression.

"You might ask, why is that? Well, studies are usually funded by a drug company that wants to make money." If you have an anti-osteoporosis drug you're developing, you'll typically pick a younger and healthier population to test it on, he explained. "Because if you pick a population of people in their eighties and nineties, there are going to be more fractures, and people are going to pass away because they're older. So there's just a higher chance that your product is disproven." McLeod let the scary moral of the story sink in. The very old are moving, in a great wave, toward a pharmacological future that is nowhere close to serving them responsibly.

Normally, with osteoporosis patients, a doctor might prescribe a bone-building drug. It can reduce the chance of a hip fracture if you fall. But Olga is not most older people. She's in little danger of tottering over. (And her hips are actually some of her strongest bones.) For Olga it's the long bones of the arms and legs you worry about—they're the ones that take the load from the jumps and sprints. And it turns out that, for complicated reasons, osteoporosis drugs may actually *weaken* the long bones. So the picture was murky.

Olga looked carefully at Dr. McLeod. He was in a position to put the brakes on her sporting career right then and there.

But he didn't. Instead he said: "Don't let anybody tell you not to do this. Nobody can quote you a study telling you it's harmful. Quite the contrary."

What he saw when he looked at Olga, McLeod said, was not the flaws—the weak spots—but the "gestalt," the whole person. "I see someone who is pretty darn healthy," he said. The lower bone density was likely the product of very gradual post-menopausal demineralization over close to fifty years. That is called life. Sports probably stemmed the decline. She *could* try drugs, but drugs can be a slippery slope with other trade-offs. She could play it safe and just quit outright, but that, too, for someone like Olga, would have consequences that are hard to measure.

"I want to compete," Olga said.

"Yup," said Dr. McLeod.

"I'm registered in three championships," she said. "The next one I leave for at the end of July."

Olga was silent for a few moments. She had been working out a plan, and she voiced it now. She would ease off the jumps at these meets. Without competition, it seemed foolish to keep pushing farther and harder; it was like bidding against yourself at an auction. "I'm not going to jump six times," Olga said. "My first clear run I'm going to stop."

"That makes sense," said Dr. McLeod.

"I've already got all of the world records in my age group," she said. Now it was just a matter of staying fit, she said, so she could charge hard at a whole new set of records when she hits 95.

Dr. McLeod nodded. "Ironically," he said, "if you were to break a bone it might have nothing to do with competing. Someone rear-ends you in the car. That's probably more likely. We all worry about the wrong thing, often. Life surprises us all the time."

Olga reached into her purse. She pulled out a gold medal from a recent meet and pushed it across the desk.

"Olga, you don't have to do this," Dr. McLeod said.

"Honestly. I can't take this. This is way too important to you for me to take this."

"Oh, go ahead," she replied. "I have six hundred of them."

In the meets that followed, Olga began telling people not to expect records. "I'm still serious in my training," she would say, "but I don't want to hit the wall."

At the BC Summer Games in Langley, British Columbia, her fifth and final major meet of the year, Olga took pains to build in more prerace rest. One afternoon she found a quiet place and lay on the grass with her feet up on a public bench. She could feel the energy flowing to her heart. She stayed that way for forty minutes. Eventually a woman came over to her and asked if she was okay. She was concerned Olga had maybe had a stroke.

On the contrary. Olga popped up like toast and, surprisingly, put in one of her best performances in years. She set new records in her sprint events. In the 200 she shattered her world record from Kamloops. In the 100 she ran faster than she had at the 2012 USA Masters Outdoor Track and Field championships in Lisle, Illinois. She was getting speedier, creeping toward the magical 23.95—the only significant world record she doesn't hold.

Things were looking stellar.

The Fall

On November 7, 2012, with the track season finished and the winter rain clouds heavy over Vancouver, Olga went out to dinner with some students she once taught in first grade—grown adults now with kids and even grandkids of their own.

The restaurant was new to her. On the way out of the complex she stopped to get her bearings. Wasn't there an elevator around here? The question was just enough of a distraction that, for a nanosecond, she stopped paying attention to where she was stepping, which happened to be on top of a steep flight of stairs. Her shoe lost its grip on the tile. The students behind watched in horror as Olga tipped forward and vanished.

The view from her own perspective, even in the moment, struck her as interesting. Diving headfirst down the stairwell, she was aware of the texture of the wall going past. She looked for something to grab but found only air. She was wearing a fairly heavy winter coat, a kind of protective shell that gave the plunge the feeling of a demonstration sport—Urban Skeleton—although "I wouldn't recommend it," she says.

The ambulance was the first she had ever ridden in.

At the hospital she overheard two doctors looking at her X-ray film. "How is this possible?" one asked the other. A 93-year-old woman takes a flying header down a dozen stairs and she does not break a single bone?

Olga hurt, though. She'd landed on her left forearm, which immediately started swelling and would bruise spectacularly. Her left hand would puff up like a ski mitt. Her forearm was a part of her body she rarely heard from. But now she could feel every tendon in there, and where they connected, and to what. Back home she put a red sponge ball, a clown's nose, between the fingers of her left hand and squeezed. The pain was bad. It was interesting. But mostly it was bad.

The fall presented a problem to be solved. "I don't fall," she has often said—and yet she had fallen. What had happened? A missed assignment in the normally flawless, unconsciously coordinated dance of body and mind? Or something simpler? Must be the new shoes, she thought. She should have taken them to the shoemaker to have the slippery soles roughed up.

There is a threshold beyond which lies old. You know you've crossed it when "fall" stops being a verb and becomes a noun.

You don't "fall"; you *have* a fall. Olga listened carefully to see if people were using that kind of language. They weren't. Because she wasn't old, right? Not yet. Old people don't bounce when they plummet down a flight of stairs.

A few months before, on our flight to Chicago, as Olga coaxed me through a Sudoku puzzle so tricky it was raising a vein in my temple, I was suddenly reminded of my grandmother. It was she who taught me to play chess. To make things fair, she would always spot me her queen. On her 100th birthday, when I was 25, she still beat me. (Though, let the record show, it was a dogfight.)

"What does it *feel like* to be a hundred?" my sister asked her afterward, taking her hands. My grandmother replied, "I hope you never find out." She was as sharp as a Cheddar, but essentially body locked. She moved, when she had to, achingly slowly, with a cane. She was like a transcendently beautiful butterfly pinned to a board. She died—was released, really—just shy of 102.

It's hard to say which is more torturous: when the body fails the mind or the mind fails the body. (Maybe the answer is the first is harder on you and the second is harder on everyone around you.) Somehow, Olga has escaped both scenarios. But, no question, minor frailties are beginning to expose themselves.

There is a creeping problem with one eye. One evening at an Italian restaurant in Kamloops, Olga observed that if she closed the other eye and looked across at her daughter Lynda, a shadow fell across Lynda's face, turning it into a crescent moon. Olga had been using eyedrops, which seemed to shrink the blind spot. She made a mental note to get her doctor to write a "therapeutic exemption use" note, in case she's drug tested.

One summer evening in 2012, driving home from a meet after dark, she realized the headlights of oncoming motorists were kind of noisy. There was a lot of light that was hard to sort out, and it was distracting. "I had to ask myself," she says, "is this safe?" Her heart was beating fairly quickly; her warning system had been engaged. She took that as a pretty clear sign. She vowed, then and there, to stop driving at night.

The previous summer, during the world outdoor championships in Sacramento, we took an afternoon walk together in the botanical garden on campus. We stopped in a clearing near some magnolias whose ancestors date back to the dinosaurs.

Olga had had a big first day, and in a typically scientific postmortem she was trying to break down the reasons. What was different? The extra garlic in the broccoli soup? The afternoon nap? The dry heat, which always seems to coax a good performance out of her?

I noticed her memory was shaky. I brought up an event that had taken place a month ago and she seemed to draw a blank. (She could still retrieve the deep past with ease; the hole in the pocket was all short-term stuff.)

Such a pattern is perfectly normal—and given my own track record here I'm in no position to judge. Still, a different order of absentmindedness was becoming apparent. Thomas Perls, head of the New England Centenarian Study, found in his Super Senior subjects who eventually showed signs of cognitive impairment that it tended to become noticeable around age 92. So I am aware now, really for the first time, of the possibility that there's more damage that we can't see.

I couldn't help wondering: these recent hairline cracks in the seeming bulletproofness were coincidental with her dialing back her effort—from the ferocious regime of Dynamite Girl to something approaching the lifestyle of a normal active senior citizen, clocking in at aquafit three mornings a week. I found myself getting frustrated with the amount that she had scaled back. Was she losing some of the protective effect of hard training?

We'll never know, and Olga makes no apologies for her choices. It's as if whatever used to compel her to train like a madwoman is slowly being undermined by something subtler and deeper.

"The psyche ripens" in old age, as the French geriatric psychiatrist Olivier de Ladoucette puts it, and as it does priorities can shift. Things that previously seemed trivial, amid the hurly-burly of the modern contest, suddenly shine with value, and things that once seemed ever so important evanesce like foam. After a lifetime of action, Olga is discovering the virtues of occasional nonaction. On a personal level, it's as if the small vulnerabilities she is becoming aware of in herself are shrinking the space between her and everybody else. It's probably making her a worse competitor but a better person.

In February 2013, Olga received a note from Harold Morioka, the legendary middle-distance masters runner turned track coach, who has been keeping tabs on the status of her world records. She now held twenty-six world records, not twenty-seven, Morioka said, because "Someone just beat your W90 shot put record." That someone is an Estonian woman named Nora Kutti. Normally word that one of her records had been poached would make Olga hungry to retake it. But Olga knows Nora, and Nora has had some personal struggles, so Olga is happy for her.

In the 200-meter final of her last international meet of 2012, Olga ran with all the competitors from age 80 on up. American Sumi Onodera-Leonard, age 84, shot out into the lead. One hundred meters in, Olga was a full five seconds behind. Then she turned it on. She almost fully closed the gap, and was close enough to touch Leonard's singlet as the pair crossed the line. After the race, Olga wondered why she didn't run faster sooner. "I think maybe part of me wanted her to win," Olga says, "because she's such a nice person."

This has been happening more and more—a little glaucoma in the eye of the tiger. Which is a higher form of respect: to ease off ever so slightly and let a deserving competitor win by a nose or honor that competitor by bringing everything you've got? She wrestles with this.

"Sometimes you have to be a human being, too," Olga says. "I thought maybe I could be a human being for just one race."

The Big C

Just before Olga was discharged from Lions Gate Hospital after her tumble down the stairs, amid talk of the mystery and marvel of her resilient bones, a doctor came over with an X-ray in hand. He asked to speak to Olga and her son-in-law, Richard. Something unexpected had turned up on the scan: a shadow halfway down her right lung, about the diameter of a silver dollar. He thought it'd be wise if a lung specialist saw the film.

In early January 2013, Olga and Lynda and Richard sat down with Tawimus Shaipanich, a respirologist at the Pacific Lung Health Centre in St. Paul's Hospital.

"How are you feeling?" he asked Olga.

"Fine," she said. "Look at me."

"No coughing?"

"No."

Dr. Shaipanich is an expert in advanced lung cancer diagnosis and early lung cancer diagnosis, which meant everyone could skip the preliminaries. The C word.

It was only by chance that everyone was seeing this picture now. Couldn't the tumor—if that was indeed what it was—have been benignly sitting there for twenty years?

The doctor allowed it was possible. The only way to know would be to take some cells. That meant a bronchoscopy—going down through her nose and grabbing a bit of tissue.

Otherwise, he explained, we wait three months and see if it's grown.

If it has? Then there were four options: surgery, radiation, chemo, or nothing.

Lynda noticed her mom's face had reddened at the mention of the procedures. She thought she detected a tear. Lynda understood. You start intervening medically and you cause more problems, "and now Mom can't compete," Lynda says—and "*that* would be the end of her."

But a few days later on the phone, Olga sounded as upbeat as I've ever heard her. Look, she said. If she had never accidentally taken an after-dinner walk that went vertically instead of horizontally, she would never have ended up in a hospital getting an X-ray that improbably revealed something she can't realistically do much about anyway. She had decided to just wait. *What's meant to be will happen.*

"Honestly, I'm not letting it worry me," she said. "Right now, if there is something there, I don't want to know about it. I'm preparing for Turin in July." A press release for those upcoming world championships had just been issued, with Olga, "one of the world's most famous masters athletes," as the face of the meet. There was even talk that her old pal Ruth Frith, now 102, would be there.

By Olga's lights, this is not nothing, this development in her lungs. But it is not yet something. She has survived worse.

"I think Mom is viewing this as just another challenge," Lynda told me not long after the consult. "Her feeling is like, 'Okay, bring it on.'"

THE human life span is just too short to learn what we need to learn. So at least a few people would need to stop aging entirely in order to gather the wisdom and perspective needed to run things in an increasingly complex world.

That is the conceit of George Bernard Shaw's play *Back to Methuselah,* where very long life emerges not as a luxury but as an evolutionary necessity. (How long does it take to become sufficiently wise? Three centuries, minimum.)

The quest for immortality was one of the first stories—at least as old as Gilgamesh—and it remains in some ways the prime story. The hard fact that nobody gets out alive is perhaps so terrifying that even the most hardheaded logicians escape into fiction, sometimes.

But there is a group of maverick scientists who don't treat it as fiction at all.

Russ Hepple remembers meeting a few years ago one of the high priests of the movement, Aubrey de Grey. The British gerontologist sat cross-legged on a table and stroked his Rasputin beard and made pronouncements like the one for which he has become most famous: "Very likely the first person who'll live to a thousand has already been born."

De Grey is chief science officer of the SENS Research Foundation, a high-profile hive of antiaging research based in California. He wants to "cure" aging. And not because it would be fun to dance at your great-great-great-great-great-great-granddaughter's wedding. He sees it as a kind of moral imperative. A lot of money and brainpower go to fighting diseases caused by aging, which account for 90 percent of all deaths in the developed world. But you could cut out the middleman if you solved aging itself.

Like a modern-day George Bernard Shaw, de Grey sees value in keeping people alive to conquer the world's thorniest problems, such as figuring out a clean-energy solution on a planet destined to get mighty crowded.

How will this godlike job get done? We'll transplant stem cells. Or we'll inject into our cells healthy mitochondria to replace the worn-out bits. Or maybe we'll create little "nanobots" to cruise around inside our bodies and repair cellular damage. The immortalists are a little vague on the details—but they swear the lock-springing knowledge is not that far away. The trick is to slow aging just enough that we're still above ground when the technological revolution delivers the answers.

Endless life holds no appeal for Olga. On the subject, she is with the British novelist Susan Ertz: "Millions long for immortality who do not know what to do with themselves on a rainy Sunday afternoon."

Once I asked her how long she would want to live, in a perfect world. Answer: 120.

She came up with the number after reading one of Earl Fee's books. "He's so fit he's going to go on and on," she said. She decided that she will live to 120 and he will live to 121.

It didn't seem polite to put it the other way around.

"I got an idea this morning," Olga announced at breakfast after the Beckman tests. "I'm thinking I'd like to donate my brain."

I put down my grapefruit spoon.

"Um . . . it feels like you're still using it," I said.

A little smile. "Ah, yes. But, you know . . ."

"Later."

"Yes."

We both sat there for a moment, letting the size of the gesture fill the silence. To Olga it just made sense. Even the best brain scan can tell us only so much. And anyway, the rest of her organs are already earmarked for dispersal and study, so why not this one? If it ends up pickled on a high shelf in a brain bank awaiting a new set of questions scientists don't yet know to ask, or traveling cross-country in the trunk of a Buick Skylark, like Einstein's? Well, then her story will go on in a different way.

But if Olga is offering herself as a model, an object of study,

I'm also becoming aware of the consequences of thinking of her that way.

These days, as we walk together I size up her gait and think about the studies linking the speed of gait to longevity. (Older folk who walk briskly have a 90 percent chance of living ten more years.) When we meet for lunch I take note of whether she's still punctual, or a minute or two late—a sign that her internal metronome is changing? When I hear her use an unusual word perfectly, I take it as a great sign—remembering how British novelist Iris Murdoch's vocabulary shrank in her later books, portending creeping dementia. I see Olga in every scrap of well-being research the way some spiritual pilgrims see God in plates of spaghetti. This is called "confirmation bias." It's a problem on almost a moral level. If I see what I expect to see in Olga when I look at her, then I am no longer really seeing her.

She is astonishing yet mortal, full of shadows. Who isn't? But what if simply learning these facts changes how she thinks of herself, even nudges her on a different course? What does it mean to Olga that her deepest beliefs are now coming under the scrutiny of science—which says there is no truth but what can be proved in double-blind studies published in peer-reviewed journals?

Some days I find myself thinking that this clock was working beautifully before we tried to understand it by taking it apart.

AND yet Olga remains committed to the terms of the deal. There is clearly value in her sacrifices. If some of her privacy has been nibbled away and her energy siphoned off, well, that's the price of sharing her message: rise up off your butts, people, and you will feel better and live longer. This has, of course, been the point all along, well before I entered the picture. It's the reason she shows up at "wellness" fairs and makes visits to nursing homes for no pay but gas fare if she's lucky. "To inspire, that's the name of the game," she repeatedly says.

The irony is, as her vulnerabilities reveal themselves, she becomes more inspiring, not less. At least to me. She is no longer quite so *other*. It used to be, when we leaned toward each other in noisy restaurants, it was because *my* hearing was failing, not hers. It used to be that while traveling *I* had to have a little catnap in her hotel room in order to keep up with her into the evening. Now, as she slows down a little and I have begun pulling out of my dive, we are meeting somewhere in the middle.

If this were a Greek myth, the exact instant of that dramatic reversal—the moment when Olga transferred her power by virtue of sharing her secrets—would here be revealed. The turning point would be that race in Sacramento. After which, upon arriving home, shamed into meaningful life change, I set Olga's list beside my list and began systematically making the two congruent, until I have erased all my bad habits.

Did it happen?

Not exactly.

I still write in the Barcalounger. Binge eating of nuts and ice cream remains an issue. I don't stretch, or self-massage, or put the phone just out of reach so I have to leap up to answer it.

But because the kids now sleep, I'm sleeping, too—and with that difference alone the worrisome memory impairment has mostly burned off. Sleep rocks. It proves almost as effective as exercise in sponging up stress, the mother of all sources of accelerated aging.

I still run, though many days running feels less like a pleasurable glide than dragging a bag of bones over speed bumps. It could be that after thirty years it's time to find a new drug-delivery system—maybe cycling. Mass recreational races called fondos are proving irresistible to recovering runners. And hey, we can always return to running after a rejuvenating hiatus. Say, around 2044.

The riddle of why I cratered and Olga still soars is by no means solved. But I'm coming to think that more answers lie above the neck than below it.

Coincident with the health issues, I was becoming someone I didn't like very much—emotionally stingy and way too uptight. I was eating without tasting, talking without sharing, exercising as if training for a prison break, failing to appreciate the fantastic privilege of getting to do a fun thing for a living, and just generally burning the goodwill of people who deserved so, so much better. Maybe our bodies are trying to tell us something when they start failing prematurely. Maybe we have business to attend to that doesn't even involve mechanics.

Olga's biggest gift to me turns out not to be a set of rules but a shift in perspective.

Look around. These are your kids. This is your wife. This is your life. Its awesomeness is eluding you. Pay attention. Yes, there will come a time when you have genuine, life-threatening ailments. But, for now, stop your kvetching. And stop dreading birthdays that end in zeros. Those zeros can pull you under, like stones in your pocket. At your age, your story is not ending: you know that.

"It's good to see you smile," my daughter Madeline said recently. I had not, by her accounting, smiled in a year. It's amazing how wrinkles disappear when you relax. Worrying about aging seems, weirdly, to accelerate the process. Let go.

Breathe.

Not long ago I asked Olga, "What do you do on the days you don't feel like going to the track?" Her expression answered the question for her. She never has those days.

"If the sun is out and the grass is dry, I'm there," she replied.

Before sweeping the long-jump runway, and fluffing up a tablecloth-sized square in the sand pit to land on, she warms up with some laps: slow and then fast.

How fast? Who knows? She doesn't keep track of her time. And I have learned to do the same.

It feels like we have all the time in the world.

CODA

Nine Rules for Living

I intend to live forever. So far, so good.

—STEVEN WRIGHT

ONE DAY IN the spring of 2011 Olga called me up. She had been asked to knock out the ceremonial first drive at a golf tournament held for and in honor of the region's mayors. Of course she said yes.

"I've already gone out and bought myself a putter," she said, enthusiastically.

"Um . . . you haven't played much golf, have you?" I said.

"Never," she said.

People assume that a jock is a jock: a raging granny on the track should be a natural from the ladies' tees. But performance is about familiarity, about brute, dogged practice—and there can't be many games more cruelly designed to expose that truth than golf. Olga was being brought in as an inspirational sideshow; I worried the inspirational part might be missing from the mix.

"Listen," I said. "Why don't we go out to the range? I'm no great shakes but maybe I can at least give you a few basic pointers."

We met on a sunny Tuesday morning. Turned out, Olga *had* done a little bit of prep. After accepting that invite to the golf tournament, she fell back on her usual MO. She went to the library. She asked for a book on golf. The librarian said, "You know, you might do better to watch a video instead." So she did.

Thankfully, it was a driver, not a putter, that she'd purchased. And she already knew how to hold it; the guy who'd sold it to her for three dollars at a yard sale had shown her.

She took quite a while to get herself properly aligned. The club was slightly too long for her, so she choked up. She took the clubhead back. Her swing was compact and upright. The club shaft never got past vertical. She punched down through the ball. It flew high and straight, with a little tour fade. It rolled right up to a junked car that was parked out there, for people to aim at—right at the 175-yard marker.

It would have been a remarkable thing in itself, even if that was the high point of the day. But Olga continued to work through her large bucket of balls. Ninety percent of them came to rest in exactly the same spot—dead center, between 150 and 175 yards. So tightly bunched were her balls out there it looked like someone had overturned a bucket of eggs.

Her eyes are failing her a bit, so the white balls disappeared to her in flight, lost against the snows of Grouse Mountain in the distance. I had to reassure her that they were good—they were all equally good.

A very few she got completely under, and they popped up and dribbled away. But I could see her recalibrating her expectations for herself. On her very last ball, at the top of her swing, she tried to find another level of strength. She was going for the fence. She brought the clubhead down fast. She whiffed. And then she laughed.

* * *

Iт was brought home to me that day that something extremely rare and fabulous had happened in this person, some convergence of luck and management. That raw hand-eye coordination was, like so much about Olga, off-the-charts abnormal. She is a black swan, a one-in-a-billion shot.

This is, on the surface of things, bad news for the rest of us, because the question everybody secretly has, the question that hovers over this whole project, is this: *Can I be like Olga?*

The short answer is probably not.

But here's the good news: Can just about all of us be *more* like Olga? For sure.

Reduced to a motto pithy enough to tattoo on a sarcopenic shoulder, one prescription that emerges from studying accomplished masters athletes like Olga is this:

BREAK A SWEAT, DAILY AND DIFFERENTLY, WITH OTHERS

Following that advice alone would make a world of difference to the quality of life of everyone over 65, in just about every way you can think of: energy, mood, cognition, libido, sleep patterns, and, yes, longevity. On one level it should be an easy sell, so copious are the studies linking exercise and social networks to happiness and health.

It's that "sweat" that's the hitch. Doesn't an easy stroll or bike ride count? Of course. But it's in the extra effort, to the perspiration point, that outsized payoffs accrue—not just in the physiological ways we have talked about, but less tangibly, too.

Philosopher Sharon Kaye, of John Carroll University in University Heights, Ohio, argues that living fully means *going for it,* physically: running and throwing things and cavorting the way our ancestors once did. "Running is your primal need, whether you are aware of it or not," she maintains. In other words, puttering in the garden is more than fine, but in order to flip the switch to "maximally alive," your workout has to generate "the adrenaline induced by a pack of hungry wolves on your tail." The fact that many seniors can actually do this— maybe not run and jump at Olga's level but move faster and exert more strenuously than people think, more primally than we ever envisioned—is one of the great recent surprises of exercise physiology.

But having interviewed dozens of masters athletes over the past four years, I've noticed other things they seem to have in common besides an appetite for effort. These are things that contribute not just to performance but to general life satisfaction. They are not all athletic things, and they're not genetically determined things. In other words, they're things you or I could do to become a little more Olga-like. Some of them are so intuitive that they only bear listing here, not explaining. Things like eat better, sleep better, breathe better, floss.

But a few habits emerged that demand fuller attention. Think of them as hard-won rules from the masters that promote vitality, longevity, and happiness. Here they are: Keep moving. Create routines (but sometimes break them). Be opportunistic. Be a mensch. Believe in something. Lighten up. Cultivate a sense of progress. Don't do it if you don't love it. Begin now.

Those are the Big Nine. At least half of them, you'll notice, involve attitude. Personality may be more or less fixed, but attitude is flexible. It can be adjusted with some simple strategies. And attitude, it turns out, matters hugely in the quality and probably even the length of our lives.

RULE ONE: Keep Moving

Nobody moves continuously, or needs to. But we do need to move *continually*. Very bad things happen when our bodies go offline for long stretches—including elevated risk of the five major killer diseases. Conversely, when we move, our bodies and brains both work better. We think faster, process information more accurately, and remember more. We lay the bricks for better, longer sleep.

RULE TWO: Create Routines
(But Sometimes Break Them)

It's not just athletes and monks and the parents of little kids who appreciate the value of routine. Routines correlate with academic success and with accomplishing things. (There is no book, you will notice, called *The Seven Ephemeral Whims of Highly Successful People*.) And they seem to matter more the older we get. Committing the more mundane parts of our life to habit and routine frees up RAM for the things that matter to us. Our bodies crave regularity.

But routines can also become ruts. "Over the years you carve yourself into a given shape," writes Robert Thurman, the Columbia University professor and Buddhist scholar. A perfectly clockwork life means no stress but also no adaptation. "The challenge," says Thurman, "is to keep discovering the green growing edge."

RULE THREE: Be Opportunistic

A classic management strategy says you should always try to figure out the 20 percent of every task that's most important, and put 80 percent of your effort there. Olga will tell you she tries her best at every competition she enters, but the results don't lie: it's at the big meets that she goes for it.

Opportunism is the reason our species is still around. Our ancestors "were active when they had to be," says William Meller, professor of evolutionary medicine at the University of California at San Francisco. But whenever they could conserve their energy, they did.

Spend your precious energy wisely, then. *A time to run, a time to hit the hot tub,* as Ecclesiastes almost said.

RULE FOUR: Be a Mensch

Kindness didn't use to have to be justified: it was an obvious virtue. Then the evolutionary psychologists got their mitts on kindness and strengthened the case for it. Doing good doesn't just feel good, it *works*. It's healthy for the tribe and healthy for us. Kind people attract the attention of others in a primitive and profound way; we are wired to recognize strangers "who may help us out in a tough situation," recent research out of Berkeley suggests.

On the Serengeti, baboons with plenty of good friends thrive. The genes of nice baboons, and nice people, go on.

RULE FIVE: Believe in Something

Believe in what? It doesn't really matter (within reason). Belief is a trait of temperament, some social scientists maintain, one marked by "the tendency to embrace puzzles, to see life's dark spots as necessary tasks," as psychologist James Fowler put it. People who look at life that way tend to thrive, by almost all measures.

Believe something, rather than nothing—while understanding that your belief may change tomorrow. Have a chip down on the table instead of in your pocket.

RULE SIX: Lighten Up

Managing stress is staggeringly important in terms of the flipping of genetic switches, the unlocking of the potential for longevity in the genome we rode in on.

But if exercise is the *best* way to shed stress, maybe the most underrated is a mental trick: the big-picture shift. Viewed from space, almost all our worries are trivial. The clock is running. There's really *no time* for grumbling—only for grace and gratitude.

RULE SEVEN: Cultivate a Sense of Progress

We need to feel as if, somehow, we're improving. That's the plain evidence of many studies of life satisfaction and human motivation. Without periodic doses of what Harvard Business School professor Teresa Amabile calls "small wins," our morale is crushed and we stop trying hard.

The solution: move the yardsticks. Adjust your expectations for yourself—just as masters track adjusts its expectations for human performance as the body ages. Then improve upon them, even by the thinnest margin. That's progress.

RULE EIGHT: Don't Do It If You Don't Love It

In his book *50 Athletes Over 50 Teach Us to Live a Strong, Healthy Life*, Don McGrath found that the single most relevant thing his elite sportspeople had in common was their sense that . . .

this is fun. The sweating, the competitive jockeying, the mild pain afterward, or even during—it's fun, all of it. It's play, not work.

If it's not fun, don't do it. That's easy—because you won't if it isn't. People can't be guilted into lasting healthy behavior change. *Should* doesn't work. Only *want to* works.

But here is the weird thing. One day you may find yourself doing something tough, maybe because you like who you're doing it with, and you may realize you don't hate it. And so you come back. That's how it starts.

RULE NINE: Begin Now

Here is the good news for boomers who lost the plot, fitness-wise, but harbor hopes of returning to form. Not only is midlife *not* too late to embark on this, providing we rev back up slowly, in some ways it's the best time to go for it. We're rested, we're restless, we're ready. Our bodies, from whom we have become estranged, are whispering to us: *Let's get back together.*

So let's.

SELECTED BIBLIOGRAPHY

Agus, David. *The End of Illness*. Free Press, 2012.

Amabile, Teresa. *The Progress Principle: Using Small Wins to Ignite Joy, Engagement, and Creativity at Work*. Harvard Business Review Press, 2012.

Ambrose, Theresa Liu. "Resistance Training and Functional Plasticity of the Aging Brain." *Neurobiological Aging*, July 2011.

Angier, Natalie. "Modern Life Suppresses an Ancient Body Rhythm." *New York Times*, March 14, 1995.

Baker, Joe, Sean Horton, and Patricia Weir, eds. *The Masters Athlete: Understanding the Role of Sport and Exercise in Optimizing Aging*. Routledge, 2010.

Barzilai, Nir. "Biological Evidence for Inheritance of Exceptional Longevity." *Mechanisms of Aging and Development*, February 2005.

Beech, Hannah. "Japan's Unlikely Saviors: Elderly Willing to Toil in a Nuke No Go Zone." *Time*, June 1, 2011.

Berglund, Christopher. *The Athlete's Way*. St. Martin's Griffin, 2008.

Bergquist, Lee. *Second Wind: The Rise of the Ageless Athlete*. Human Kinetics, 2009.

Booth, Frank et al. "Waging War on Modern Chronic Diseases." *Journal of Applied Physiology*, February 2000.

Bortz, Walter. *Next Medicine: The Science and Civics of Health*. Oxford University Press, 2011.

Brady, Catherine. *Elizabeth Blackburn and the Story of Telomeres*. MIT Press, 2009.

Bronson, Po, and Ashley Merryman. "Why Some Kids Can Handle Pressure While Others Fall Apart." *New York Times Magazine*, February 6, 2013.

Buettner, Dan. *The Blue Zones: Lessons for Living Longer from the People Who've Lived the Longest*. National Geographic, 2010.

Caulfield, Timothy. *The Cure for Everything*. Viking Canada, 2011.

Chaddock, Laura. "Are Higher-Fit Children Better Street-Crossing Multi-taskers?" *Journal of the American College of Sports Medicine*, March 2011.

Coates, John. *The Hour Between Dog and Wolf: Risk Taking, Gut Feelings, and the Biology of Boom and Bust*. Random House Canada, 2012.

Costa, P. T., and R. R. McRae. "The NEO-PI-3: A More Readable Revised NEO Personality Inventory." *Journal of Personality Assessment*, 2005.

Craig, A. D. (Bud). "Interoception and Emotion: A Neuroanatomical Perspective." In *Handbook of Emotions*, 3rd ed. Edited by Michael Lewis, Jeannette M. Haviland-Jones, and Lisa Feldman Barrett. Guilford Press, 2008.

Crowley, Chris, and Henry Lodge. *Younger Next Year*. Workman, 2007.

Damisch, Lysann et al. "Keep Your Fingers Crossed! How Superstition Improves Performance." *Psychological Science*, June 2010.

de Grey, Aubrey. *Ending Aging: The Regeneration Breakthroughs That Could Reverse Human Aging in Our Lifetime*. St. Martin's Press, 2008.

de Ladoucette, Olivier. *Rester jeune, c'est dans la tête*. Odile Jacob, 2005.

Dement, William, and Christopher Vaughan. *The Promise of Sleep: The Scientific Connection Between Health, Happiness, and a Good Night's Sleep*. Pan Macmillan, 2001.

Doidge, Norman. *The Brain That Changes Itself: Stories of Personal Triumph from the Frontiers of Brain Science*. Viking Penguin, 2007.

Dweck, Carol. *Mindset: The New Psychology of Success*. Ballantine, 2008.

Ekirch, Roger. *At Day's Close: Night in Times Past*. W. W. Norton, 2006.

Fee, Earl. *One Hundred Years Young the Natural Way*. Trafford Publishing, 2011.

Friedman, Howard, and Leslie Martin. *The Longevity Project: Surprising Discoveries for Health and Long Life from the Landmark Eight-Decade Study*. Penguin, 2011.

Haidt, Jonathan. *The Happiness Hypothesis: Finding Modern Truth in Ancient Wisdom*. Basic Books, 2006.

Hennezel, Marie de. *The Warmth of Your Heart Prevents Your Body from Rusting*. Pan Macmillan, 2011.

Holmes, Hannah. *Quirk: Brain Science Makes Sense of Your Peculiar Personality*. Random House, 2011.

Hutchinson, Alex. *Which Comes First: Cardio or Weights? Fitness Myths, Training Truths, and Other Surprising Discoveries from the Science of Exercise*. William Morrow Paperbacks, 2011.

Jerome, John. *The Elements of Effort: Reflections on the Art and Science of Running*. Breakaway Books, 1999.

Kaye, Sharon. "The Running Life: Getting in Touch with Your Inner Hunter-Gatherer." In Michael W. Austin, ed., *Running and Philosophy: A Marathon for the Mind*. Blackwell, 2007.

Kramer, Arthur et al. "Aging, Fitness and Neurocognitive Function." *Nature,* July 1999.

Kramer, Arthur, and Stanley Colcombe. "Cardiovascular Fitness, Cortical Plasticity and Aging." *Proceedings of the National Academy of Science of the United States of America,* February 20, 2004.

Langer, Ellen J. *Counterclockwise: Mindful Health and the Power of Possibility*. Ballantine, 2009.

Levy, Becca. "Mind Matters: Cognitive and Physical Effects of Aging Self-Stereotypes." *Journals of Gerontology,* July 2003.

Lieberman, Daniel, and Dennis Bramble. "Endurance Running and the Evolution of Homo." *Nature,* November 18, 2004.

Mah, Cheri D. "The Effects of Sleep Extension on the Athletic Performance of Collegiate Basketball Players." *Sleep,* July 2011.

Marlowe, Frank. *The Hadza: Hunter-Gatherers of Tanzania*. University of California Press, 2010.

Marmot, M. G. et al. "Health Inequalities Among British Civil Servants—the Whitehall II Study." *Lancet,* June 1991.

McDougall, Christopher. *Born to Run*. Random House, 2009.

Meller, William. *Evolution Rx: A Practical Guide to Harnessing Our Innate Capacity for Health and Healing*. Perigee Trade, 2009.

Noakes, Timothy. *Lore of Running*, 4th ed. Human Kinetics, 2002.

O'Keefe, James. "Organic Fitness: Physical Activity Consistent with Our Hunter-Gatherer Heritage." *Physician and Sportsmedicine,* December 2010.

Parker, John L., Jr. *Once a Runner*. Scribner, 2010.

Perls, Thomas, and Paola Sebastiani. "The Genetics of Extreme Longevity: Lessons from the New England Centenarian Study." *Frontiers in Genetics,* November 2012.

Pillemer, Karl. *30 Lessons for Living: Tried and True Advice from the Wisest Americans.* Hudson Street Press, 2011.

Ramlakhan, Nerina. *Tired but Wired.* Souvenir Press, 2010.

Ratey, John. *Spark: The Revolutionary New Science of Exercise and the Brain.* Little, Brown, 2008.

Reynolds, Gretchen. *The First 20 Minutes: Surprising Science Reveals How We Can Exercise Better, Train Smarter, and Live Longer.* Plume, 2013.

Rockwood, Kenneth et al. "A Global Clinical Measure of Fitness and Frailty in Elderly People." *Canadian Medical Association Journal,* August 30, 2005.

Sansonetti, Ugo. *Never Stop.* Sports Society Press, 2003.

Sapolsky, Robert. *Why Zebras Don't Get Ulcers,* 3rd ed. Times Books, 2004.

Schleip, Robert. "Fascial Plasticity: A New Biological Explanation." *Journal of Bodywork and Movement Therapies,* January 2003.

Shaw, George Bernard. *Back to Methuselah.* Project Gutenberg, 2004.

Shields, David. *The Thing About Life Is One Day You'll Be Dead.* Vintage, 2009.

Snowden, David. *Aging with Grace: What the Nun Study Teaches Us About Leading Longer, Healthier, and More Meaningful Lives.* Bantam, 2002.

Stern, Yaakov. "Cognitive Reserve." *Neuropsychologia,* March 2009.

Stevens, John. *The Marathon Monks of Mt. Heie.* Echo Point, 2013.

Tarnopolsky, Mark et al. "Endurance Exercise Rescues Progeroid Aging and Induces Systematic Mitochondrial Rejuvenation in mtDNA Mutator Mice." *Proceedings of the National Academy of Science,* 2011.

Thompson, Kevin. "Pushing the Limits of Performance." *Science Daily,* October 17, 2011.

Vaillant, George. *Adaptation to Life.* Harvard University Press, 1998.

————. *Triumphs of Experience: The Men of the Harvard Grant Study.* Belknap Press, 2012.

Vernikos, Joan. *Sitting Kills, Moving Heals: How Everyday Movement Will Prevent Pain, Illness, and Early Death.* Linden Publishing, 2011.

ACKNOWLEDGMENTS

I owe a tremendous debt to many people.

Sam Stoloff, for ongoing encouragement and tactical precision.

Gillian Blake at Henry Holt and Craig Pyette at Random House of Canada, for believing in the story and helping to improve it.

Ken Stone, the true Boswell of the masters track community, for contacts and context.

Michele Coviello, for translation help and invaluable perspective.

Ilena Silverman, who made the first bet.

Harold Morioka, for his meticulous record keeping.

Curtis Gillespie and Caroline Zancan for helpful feedback on early drafts.

William Meller and Paul Ingraham, who made me question everything I thought I knew. ·

Nancy South and Iain Mant, for a timely sanctuary.

Chris O'Connell, for his sharp eyes.

Tamotsu Nagata, for diplomatic outreach.

Michael Barrand and Bill McNamara for shoulders to lean on.

My wife, Jen, who, as ever, belayed so I could climb.

My sisters, Carol, Wendy, and Lynn, who made a crucial decision that helped this book happen.

And not least, Olga herself, for saying yes.

INDEX

BRUCE GRIERSON is the author of *U-Turn: What If You Wake Up One Morning and Realized You Were Living the Wrong Life?* He is an award-winning freelance writer whose work has appeared in *The New York Times Magazine, Psychology Today,* and *Scientific American,* among other publications. He lives in North Vancouver, Canada.